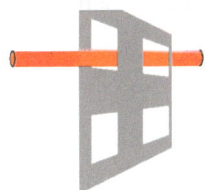

SEEING THE WORLD

Richard A. Clement

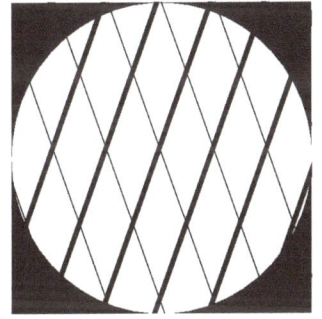

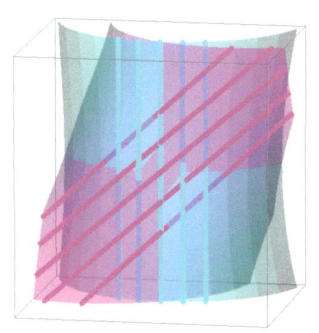

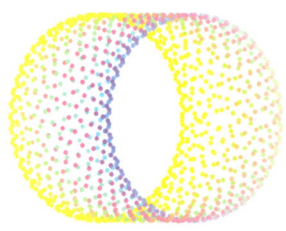

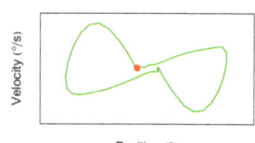

Technical note: All the figures were created using Mathematica® Version 6.0.

Indexed by Christine Boylan

© Richard A. Clement 2008

ISBN 978-0-9558515-0-6

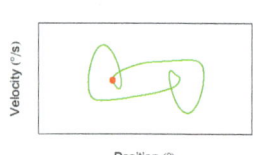

CONTENTS

1 A folded piece of card

1.1 Is another book on vision necessary?
Books on vision - A striking phenomenon

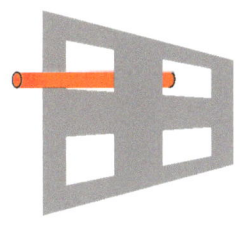

1.2 Physiological optics
Unconscious inference - Cubic corners - Reversible figures

1.3 Gestalt theory
Phi motion - Minimum energy configuration

1.4 Ecological optics
Optic array - Extraction of invariants - Effect of head movement

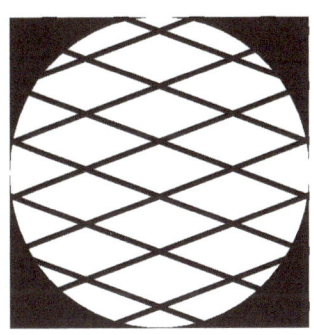

1.5 Computational vision
Levels of explanation - Mutual reflections

1.6 Visual Neuroscience
Multiple visual areas - Functional specification

1.7 The mechanism of seeing
Questions

2 Up with geometry and down with algebra in all things visual

2.1 Algebra and geometry
Vector geometry - Algebraic geometry - Linear neurons

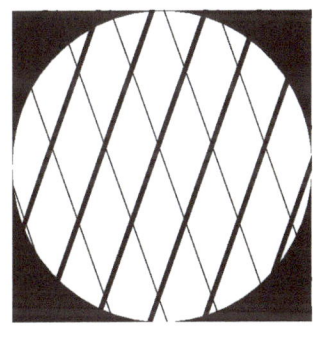

2.2 Motion and stereoscopic parallax
Viewpoint invariance - Aperture problem - Correspondence problem

2.3 Eye movement coding
Superior colliculus

3. Seeing things

3.1 Reconstruction, reconstruction, reconstruction
Hermann grid - Receptive field

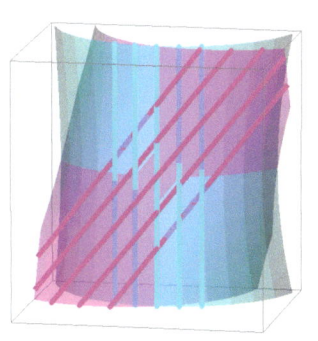

3.2 Informative features
Cortical receptive fields – Unresponsive regions - Tracings

3.3 Eye movements and working memory
Fixation eye movements – Build-up of information

4 Seeing space

4.1 Dimensionality reduction
Decorrelation - Munsell Chip reflectance functions

4.2 Colour vision
Cones - Opponent colour channels

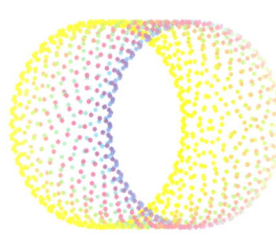

	4.3	Spatial vision
		Spatial frequency channels - Opponent spatial channels
	4.4	Eye rotations
		Euler angles - Listing's law
	4.5	Visual imagination
		Tesseract - Plane diagrams - 4D embeddings
	5	**Against block diagrams**
	5.1	Structure and algebra
		Sensory transformations - Motor transformations
	5.2	Behaviour and geometry
		Slow-fast systems - Dynamics of nystagmus
	5.3	Albinism and the visual system
		Underdeveloped fovea - Misrouting
	5.4	How does the brain see?
		Summary

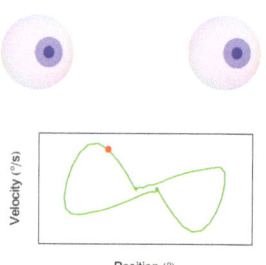

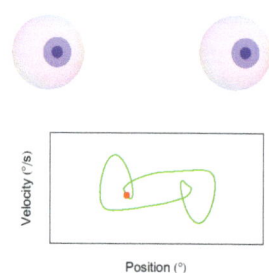

Flip books

There are ten flip books incorporated into the text. The first five are printed on the odd-numbered pages and the second five are printed on the even-numbered pages. The second five run from the back to the front of the book. Starting at the top of the page and reading downwards these are as follows:

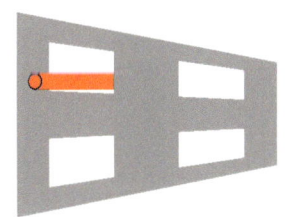

Ames window: A rotating trapezoidal shape appears to oscillate to and fro. See page 22.

Aperture 1: Two sets of crossed lines appear to move together vertically. See page 24.

Aperture 2: Two sets of crossed lines appear to move over each other. See page 24.

Line horopters: The positions of various line horopters. See page 25.

Object tracking: Two overlapping stimuli which change form and yet remain distinguishable. See page 64.

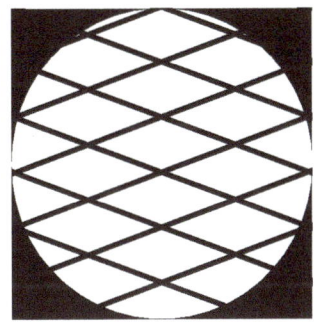

Torus: A four dimensional representation of a torus. See page 53.

Projective plane: A four dimensional representation of a projective plane. See page 54.

Dynamics of nystagmus 1: The dynamics of jerk nystagmus. See page 62.

Dynamics of nystagmus 2: The dynamics of pendular nystagmus. See page 62.

Biological motion: A example of a biological motion stimulus. See page 64.

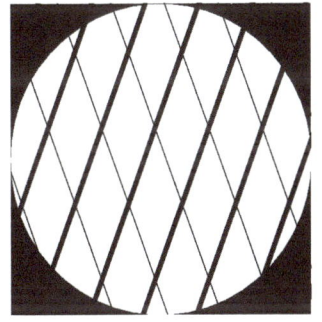

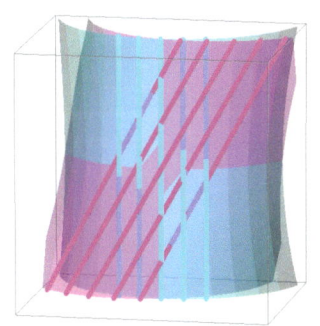

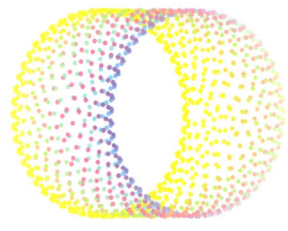

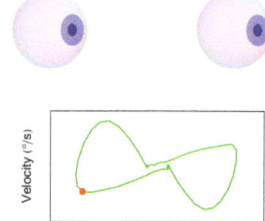

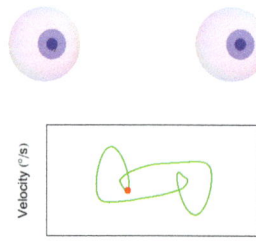

1.0 A FOLDED PIECE OF CARD

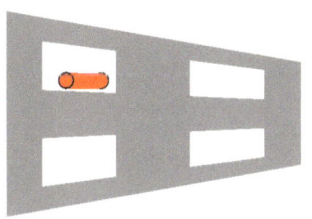

1.1 Is another book about vision necessary?

In the last few decades books on vision have been published regularly at the rate of more than one a year. Table 1 lists some of the books of vision published during the last decade. Each one contains the distillation of many years of thoughts on the subject. So surely we should be able to draw a line under the subject by now?

Table 1 Some of the books on vision published during the last decade.

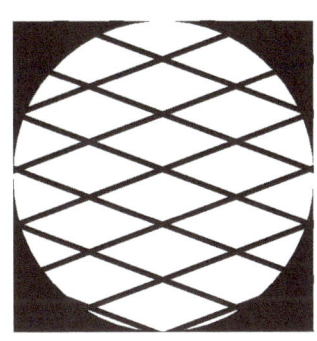

1998	First Steps in Seeing
	R.W.Rodieck, Sinauer Associates, Inc., Sunderland
1998	La Science des Illusions
	J.Ninio, Editions Odile Jacob, Paris
1998	Visual Intelligence
	D.D.Hoffman, W.W.Norton & Company, Inc., New York
1999	Sight Unseen
	G.Kleege, Yale University Press, New Haven
2000	Human Perception of Objects
	D.Regan, Sinauer Associates, Inc., Sunderland
2001	Computational Neuroscience of Vision
	E.T.Rolls and G.Deco, OUP, Oxford
2003	Active Vision
	J.M.Findlay and I.D.Gilchrist, OUP, Oxford
2003	The Space Between Our Ears
	M.Morgan, Weidenfeld & Nicolson, London
2003	Why We See What We Do
	D.Purves and R.B.Lotto, Sinauer Associates, Inc., Sunderland
2003	Seeing and Visualizing
	Z.Pylyshyn, MIT Press, Cambridge
2004	The Quest for Consciousness: A Neurobiological Approach
	C.Koch, Roberts and Co., Englewood
2004	Sight Unseen
	M.A. Goodale and A.D.Milner, OUP, Oxford
2004	The Thinking Eye, The Seeing Brain
	J.T.Enns, W.W. Norton and Company, New York
2006	Seeing Red
	N.Humphrey, Harvard University Press, Cambridge
2006	Seen and Unseen: Art and Science from a Different Perspective

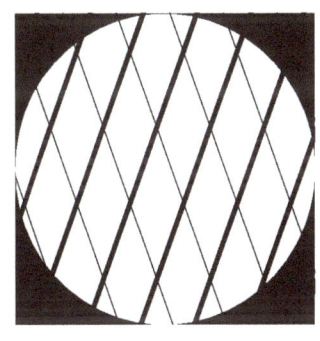

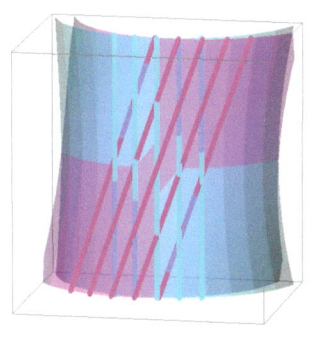

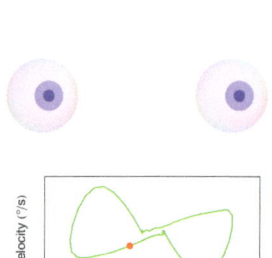

M.Kemp, OUP, Oxford

2006   Basic Vision: An Introduction to Visual Perception.
R.Snowden, P.Thompson and T.Troscianko, OUP, Oxford

Unfortunately not. The trouble is that we are still hard pushed to explain any perceptual phenomena in terms of known physiological mechanisms. For example, take a business card, fold it lengthways and place it on a table with the fold upwards. View the card along the fold with one eye closed, and you will see a sudden change in the shape of the card, so that it appears to be standing upright on the table, with the fold furthest away [1]. If you have not seen this phenomenon for yourself, then it is worth persevering as it can take several minutes before it appears for the first time, but the apparent reversal of a solid object is very dramatic when it occurs.

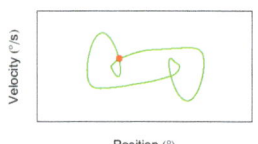

The difficulty in explaining this phenomenon arises from there being just two interpretations of a single retinal image when an unlimited number of scenes could have produced the image [2], as illustrated in figure 1.1. Now, a folded piece of card is, by anybody's standards, one of the simpler visual stimuli, so any half decent theory should have no difficulty explaining the visual phenomena associated with the card. What I am going to do is illustrate different approaches to explaining vision, using the appearance of a folded piece of card as a test case.

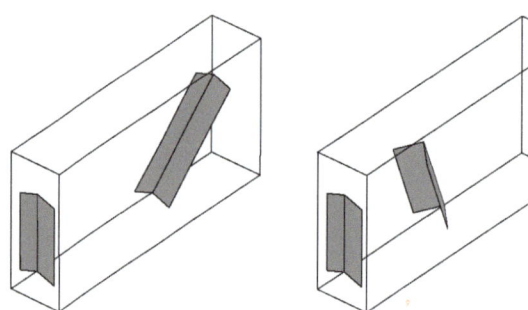

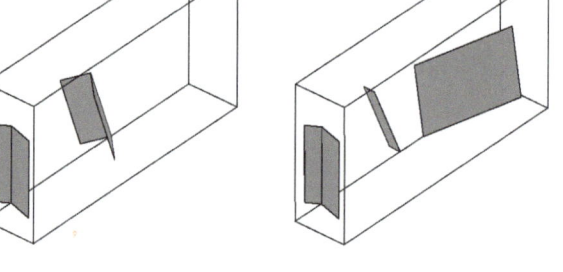

Figure 1.1. Illustration of just 3 of the possible 3D scenes which all project onto the same retinal image of a folded piece of card. As well as a boundless number of possible scenes with a convex fold and with a concave fold, there are also an infinity of scenes in which the two halves of the card are not even connected.

1.2 Physiological optics

A logical approach to understanding vision is to investigate the successive transformations by which the visual system turns light from an object into perception of the object. This is the approach of physiological optics, which is comprised of three parts: the theory of the path of light in the eye, the theory of the sensations of the neural mechanism and the theory of the interpretation of the visual sensations in terms of objects. These three divisions roughly correspond to the transformations which occur in the eye and the visual cortex, as illustrated in figure 1.2. Simple observation shows that what one sees is the object that would have to be present to produce the given retinal stimulus. For example, if I press on the right hand side of my lower eyelid, then I see a bright ring in the upper left of my visual field, which is where the light would have to be to stimulate the part of the retina which is being pressed upon. This aspect of vision can be explained by assuming that sensations of light and colour are tokens, the meaning of which are determined by intellectual judgements. This process is analogous to logical deduction, except that we are not conscious of it, so that the process of seeing can be considered to be one of unconscious inference [3].

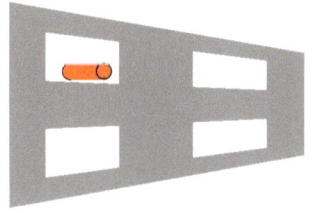

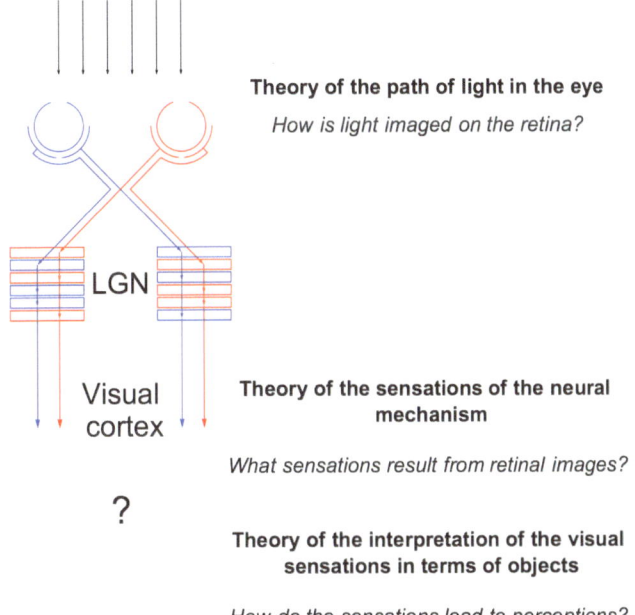

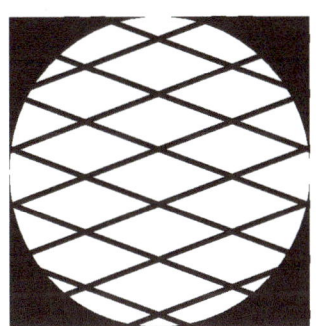

Figure 1.2 Schematic illustration of how the transformation of light into seen objects is mapped onto the neural circuits responsible for vision.

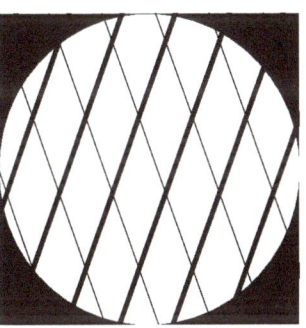

It is plausible that the rules which the visual system uses to make unconscious inferences are learned from experience. For instance, the everyday world is full of carpentered objects which have cubic corners that are made up of the intersection of three mutually perpendicular surfaces, so it is not unreasonable to assume that angles on the retina are projections of right angles. But humans do better than that, in that they are adept at judging when it correct to assume that an image could from a cubic corner [4]. The rule is, that if three lines which join in a Y shape in a picture represent a cubic corner, then none of the angles in the picture can be smaller than a right angle. Figure 1.3 shows examples of pictures of boxes which could and could not have cubic corners. In this framework, the apparent shape of the folded piece of card is inferred from its likeness to a part of a solid block.

As well as explaining the apparent shape of the folded card, the physiological optics approach can also be used to explain the apparent reversal of the card. According to the theory, what one sees is a conclusion drawn by unconcious inference from one's sensations, and in some cases more than one conclusion may be tenable [5]. This phenomenon is frequently seen in line drawings of crystals, such as that shown in figure 1.4 [6], which appears to reverse in depth, and the line drawing of the folded card can be considered as another example where more than one inference fits the data.

Figure 1.3 Pictures of boxes. On the left is a picture which could be of a box with cubic corners. On the right is a picture which could not be of a box with cubic corners.

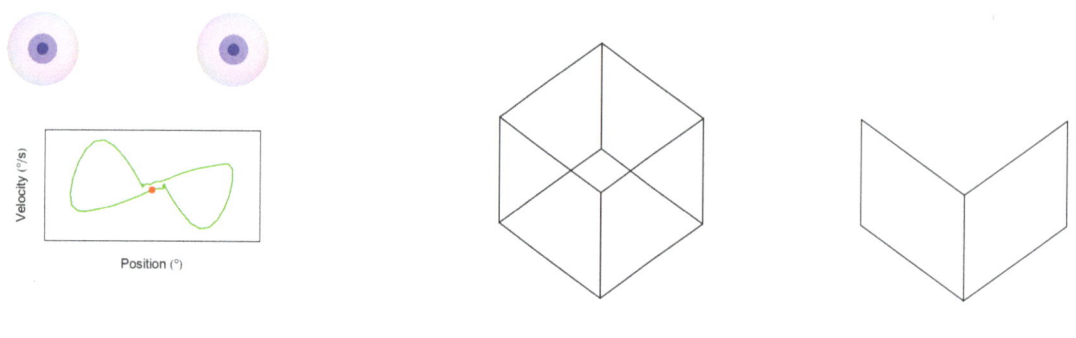

Figure 1.4 Examples of figures which reverse in depth. A line drawing of a cubic crystal is shown on the left and of the folded card on the right.

The drawback of the physiological optics approach is that it does not specify the mechanism responsible for unconscious inference, so that it gives no insight into the situations where we do not see the object which would have to be present to produce the given pattern of retinal stimulation. Take the interpretation of shadows. If one rests a pencil on a folded piece of card then the card still reverses when viewed monocularly. Yet if one could interpret the shadow of the pencil correctly, this reversal would not be possible because when the shadow touches the pencil, the fold must be convex [7].

The idea that we see the object which would have to be present to produce a given image can be given a more precise formulation by considering the probabilities involved. Let p(O) be the likelihood of the object, p(I) be the likelihood of the image and p(I|O) the likelihood of the image given a particular object, then the probability of there being a particular object O present given an image I can be calculated by Bayes's formula:

$$p(O|I) = (p(I|O) \times p(O))/p(I)$$

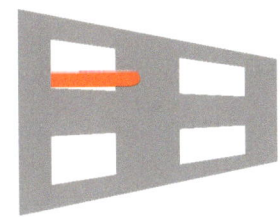

Whilst this formula is very useful or calculating what an ideal observer will 'see' when the possible objects and images are known and the probabilities can be calculated, its wider application rests on the assumption that the brain knows what possible objects are out there without giving insight into how the brain does this.

1.3 Gestalt theory

Underlying the physiological optics approach to vision is the idea that each nerve fibre has its own 'specific energy' which gives rise to an associated sensation, and that the sensations are subsequently interpreted by an unconscious inference mechanism. An implication of the approach is that objects are represented by a mosaic of firing of nerve cells terminating in the cortex and it is hard to explain within this framework why we see distinct wholes - a tree here, a car there [8].

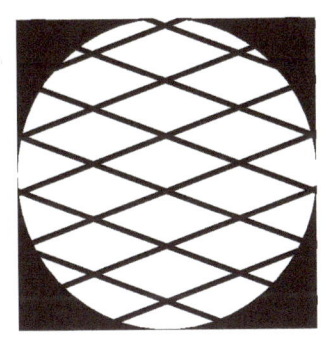

The approach of gestalt psychology was to assume that the neural processes are not independent. As such, the neural processes have properties, referred to as gestalt properties, which are more than just the sum of the individual processes. An example of such a property is provided by the phenomenon of phi movement. If two spatially separated lights are switched on and off in sequence so that one is switched off as the other is switched on, then at low rates of switching that is what one sees, but at higher rates of switching, a percept of movement is seen, referred to as phi movement.

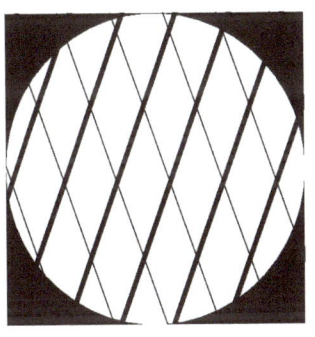

According to the approach of gestalt psychology, the neural mechanism underlying perception can be thought of as an analogue system, which changes state in accordance with the geometrical constraints on image formation, but which settles into a minimum energy level configuration corresponding to the simplest possible percept. The dynamical switching from one stable configuration to another certainly corresponds to how we see. For instance, if the top half of the folded card is sliced off, as illustrated in figure 1.5, the reversed figure appears asymmetrical, with the lower angles appearing as right angles, but if the card is tilted so the image angles become more similar it suddenly switches to looking symmetrical, but not rectangular. However, the neural mechanism responsible for such switching remains to be specified.

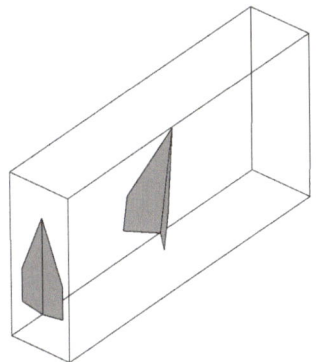 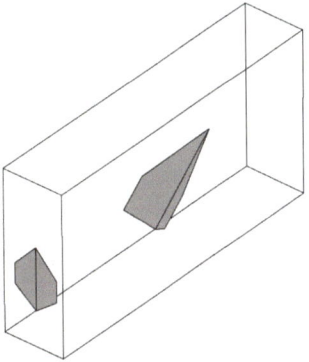

1.5 Projection of an asymmetric folded card. The differences between the reversed and the actual shape of the card become more striking as the projection becomes more nearly mirror symmetric about a horizontal axis.

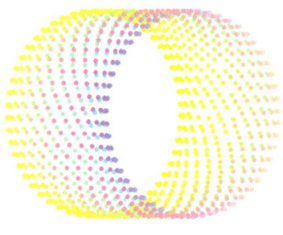

1.4 Ecological optics

Another aspect of the physiological optics approach to vision is that it gives rise to three main questions: how is light imaged on the retina, what sensations result from the retinal images and how do these sensations lead to perceptions? The approach provided a way of extending the known physics about the paths of light rays through optical system to an understanding of vision, but is misleading because it emphasises the static pictures on the retina. If one assumes that the basis of visual perception is a sequence of snapshots, then this approach leads on to the question as to how such snapshots are integrated in memory to obtain a subjective visual world. If instead, one assumes that there are invariants in the scene which transcend any particular retinal image, then there is no need to invoke memory, as awareness of the environment is directly perceived [9].

In contrast to the physiological optics approach, the ecological optics approach to vision is concerned the relation between light energy and the environment, which enables human beings and animals to move about the world. Following from the ecological optics approach, one can consider the intensity of light arriving from different directions for a particular position of an eye. The pattern of intensities is referred to as the optic array. For a moving animal, the transformation of the patterns in the array will contain information about the objects in the environment, and vision should be explained in terms of higher order variables in the optic array.

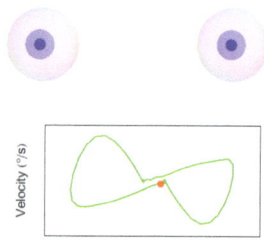

But the weakness of the ecological optics approach has always been in the actual specification of the higher order invariants of the optical array. For example, if one moves ones head from side to side while viewing the folded card then the changes in the optic array are the same, whether or not the card appears reversed, as shown in figure 1.6. But, when one moves ones head when the card appears as it is, then it does not change its shape, but when one moves ones head with the card in its reversed shape, then it appears to twist and deform [10]. Hence the perception of the card does not depend directly on invariants of the optic array.

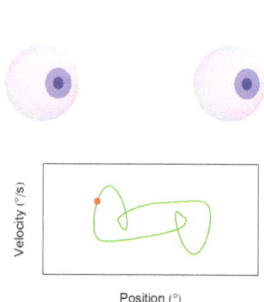

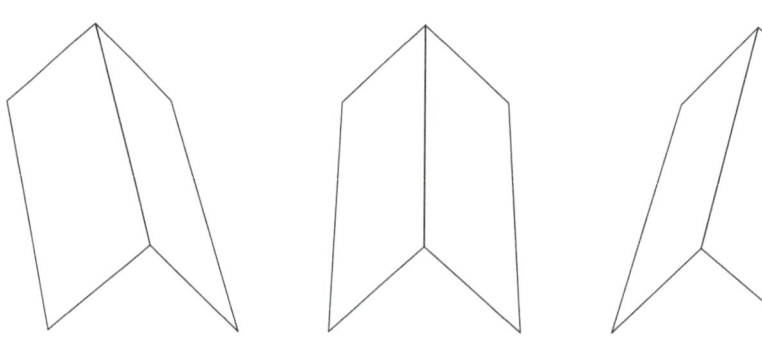

Figure 1.6 Changes in the retinal image of the folded card with side-to-side head movement.

1.5 Computational vision

A new approach to understanding vision came with the development of computers and cameras that could be connected together and used to process images of everyday scenes. Consideration of the behaviour of these systems raised the question of what is involved in making a machine that can see. This led to the realisation that the human visual system effortlessly performs many visual tasks which it turns out to be extremely difficult to make a machine to do. The computational approach to vision arose from the need to make explicit the representations used in vision and how they are manipulated. A distinction can be made between investigation of vision at the computational level, the psychological level and the neural level [11]. The computational level involves isolating the physical constraints involved in solving a perceptual task and devising a procedure for solving the task, independent of mechanism.

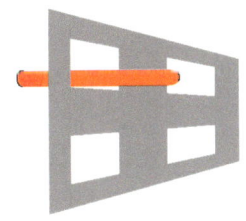

In the case of a rectangular piece of folded card one physical constraint is that edges which form an acute angle in a picture plane through the corner lie on opposite sides of the plane, whereas edges at an obtuse angle lie on the same side of the plane [1]. To make use of this constraint, draw an arrow pointing from the near end to the far end on each line segment in the retinal image corresponding to an edge. If two line segments join at an acute angle then the end labels at the join must be different (either head to tail or tail to head), and if two line segments join at an obtuse angle then the end labels at the join must be the same (either head to head or tail to tail) [12]. Starting with a guess at one line segment one can work round the entire image to obtain a consistent labelling as illustrated in figure 1.7. This scheme generates perceptual hypotheses in keeping with subjective experience as it produces a consistent labelling for the cover figure which tends to look rectangular despite this being physically impossible[13]. It also fails to generate a consistent labelling for the box which could not be cubic in figure 1.3.

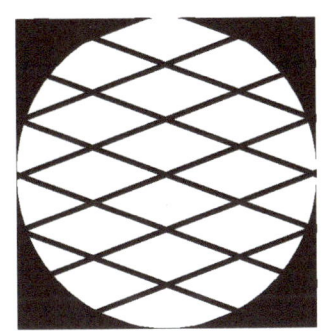

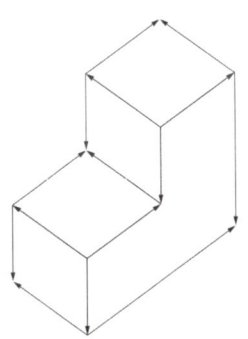
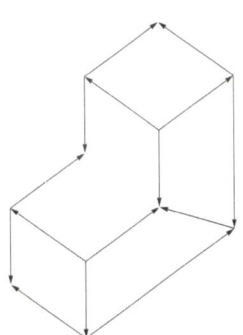
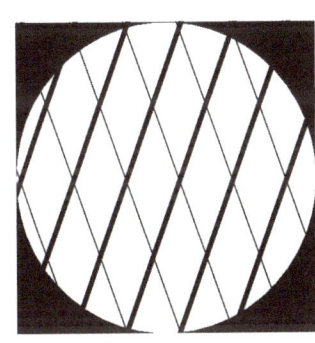

Figure 1.7 Examples of depth labeling. Each arrow points towards the end of the line which is further away.

A more sophisticated example of a computational task is provided by the interpretation of scenes which include the effects of inter-reflections between objects. The computational task involves devising a procedure to calculate the mutual illumination effects. Subsequently one can test at the psychological level if the computational procedure is being used. If one half of a concave folded piece of card is painted magenta, then the reflected illumination is taken into account by the visual system and the unpainted half of the card appears predominantly white. However, as soon as the card reverses, and mutual illumination can no longer be occurring, the unpainted half appears pink [14].

Like some of the other theories of vision, the computational approach has delivered less than was initially promised. Although it is clear that part of understanding the visual system will involve identifying the type of representations that the brain uses, this does not necessarily mean that the computational task can be analysed independently of its physical implementation. Rather it appears that the sort of computation that the visual system can carry out is determined by the nature of the

mechanism of the visual system [15].

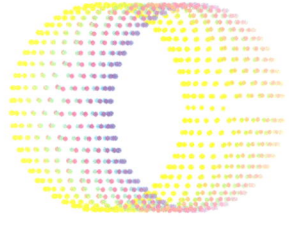

1.6 Visual Neuroscience

The early stages of the visual pathway are topographically organized in that the spatial order in the neural representations of objects corresponds to the spatial order of objects in the scene. In the cortex, visual area 1 (V1) is the largest visual area, extending over 2250 mm in humans. Bordering V1 is V2, which is in turn bordered by V3, as shown in figure 1.8. Whilst V1 forms a continuous representation of the visual field, the representation of V2 is split into an upper and a lower half. The boundary of V1 and V2 is formed by the vertical meridian and the two halves of V2 can be folded onto V1 so that they meet along the representation of the horizontal meridian. This arrangement minimises the lengths of the connections between corresponding positions in the visual field in V1 and V2.

The later stages of the visual pathway can be divided into a dorsal and ventral pathways. The dorsal pathway leads from the motion-specialised areas of cortex Medial Temporal (MT) and Medial Surperior Temporal (MST) to the parietal lobe and is responsible for processing changing spatial relationships as one moves around the world. The ventral pathway leads from V4 to the inferotemporal cortex and is responsible for recognition [21]

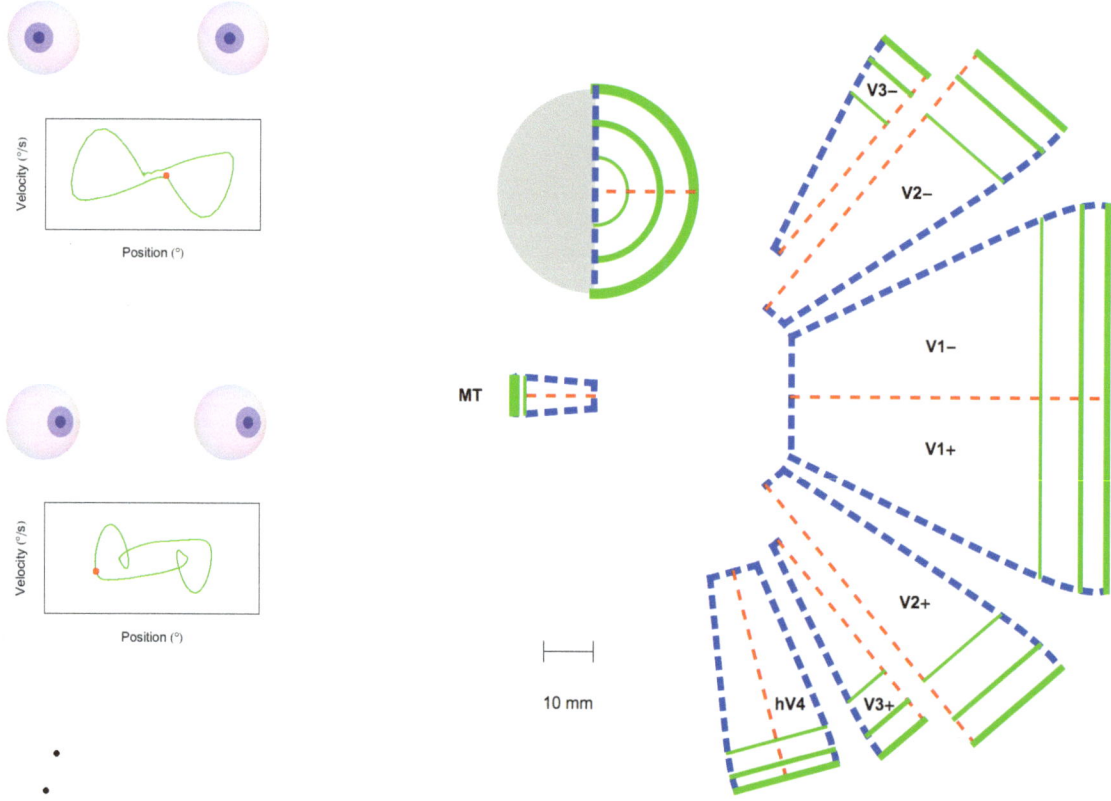

Figure 1.8 Simplified representation of some of the topographic maps in the cortex, derived from neurophysiological investigations [16-20]

1.7 The mechanism of seeing

All these theories are valid as far as they go. Whatever mechanism is responsible for vision will have to be capable of answering the big questions posed by the different theories: how does the visual system guess what would have to be 'out there' to produce the retinal image, how does the visual system isolate individual objects in the scene, how are vision and movement combined and what is all the neural activity in the different cortical maps representing?

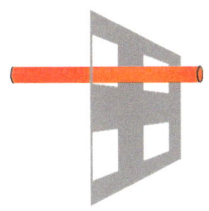

At one level it is obvious what the mechanism of vision is - a set of inter-connected neurons with sensory inputs and motor outputs as shown in figure 1.9, but does this tell us anything about what to expect from the neural mechanism of vision - and especially how it can go wrong? There is an apocryphal story about the mathematician Hardy, who wrote an equation on the blackboard and said "The proof of this is obvious". After a little while he murmured "At least I think it is", and left to check the proof in the library, only returning just before the end of the lecture to announce "Yes, it is obvious".

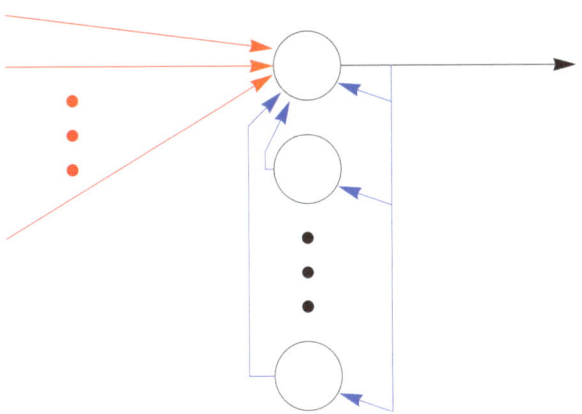

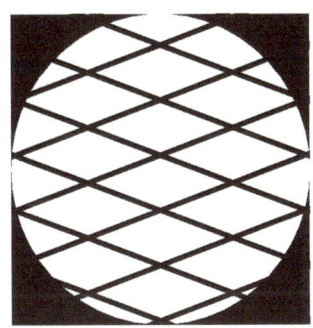

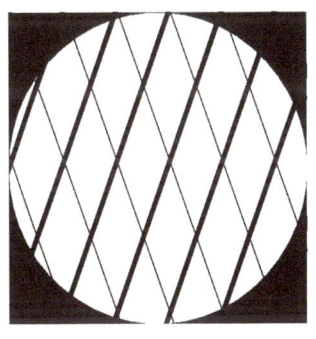

Figure 1.9. Schematic diagram of the connections of a neuron in a network. Feedforward connections are shown in red, and feedback connections are shown in blue.

1.8 References

1] Mach E The Analysis of Sensations. 1886. Translated by C.M.Williams and S.Waterlow, New York: Dover, 1959.

2] Clement RA An application of the gradient space representation. Perception & Psychophysics 1986, 39, 222-224.

3] Helmholtz H von Treatise on Physiological Optics. 1910. Translated by J.P.C.Southall, New York, Dover, 1924.

4] Perkins DN Visual discrimination between rectangular and nonrectangular parallelpipeds. Perception & Psychophysics 1972, 12, 396-400.

5] Gregory RL Eye and Brain. Weidenfeld and Nicolson, London., 1966.

6] Necker LA Observations on some remarkable optical phenomena seen in Switzerland and an optical phenomena which occurs on viewing a figure of a crystal or geometrical solid. Philosophical Magazine and Journal of Science. Third series. 1832, vol 1. 329 - 337.

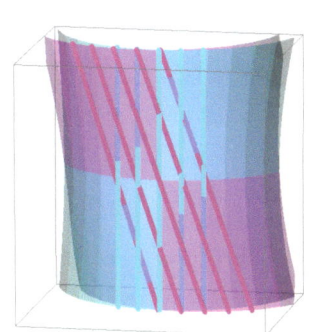

7] Mamassian P, Knill DC and Kersten D The perception of cast shadows. Trends in Cognitive Sciences 1998, 2, 288-295.

8] Köhler W. Physical Gestalten. Selection 3, pp 17 - 54,1938, in "A Source Book Of Gestalt Psychology". Edited by W.D.Ellis, London, Kegan Paul, Trench, Trubner & Co. Ltd.

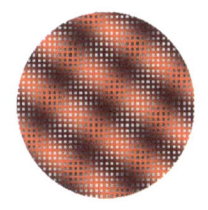

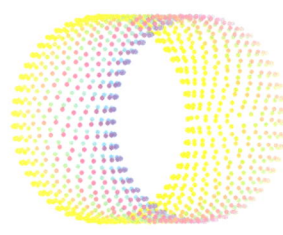

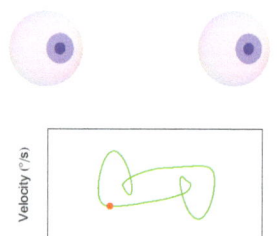

9] Gibson JJ The Ecological Approach to Visual Perception. Lawrence Erlbaum Associates, Hillsdale, New Jersey, 1986.

10] Eden M A three-dimensional optical illusion. MIT Research Laboratory of Electronics Quarterly Progress Report 1962, 64, 267-274.

11] Marr D Vision. W.H.Freeman and Company, New York, 1982.

12] Cowie R Impossible objects and the things we do first in vision. British Journal of Psychology 1988, 79, 321-338.

13] Mackworth AK Interpreting pictures of polyhedral scenes Artificial Intelligence 1973, 4, 121-137.

14] Bloj MG, Kersten D, and Hurlbert AC Perception of three-dimensional shape influences colour perception through mutual illumination. Nature 1999, 402, 877-879.

15] Morgan MJ Computational theories of vision (review of Marr). Quarterley Journal of Experimental Psychology, 1984, 36A, 157-165.

16] Gattass R, Nascimento-Silva S, Soares J.G.M, Lima B, Jansen AK, Dioga ACM, Farias MF, Marcondes M, Botelho EP, Mariami OS, Azzi J and Fiorani M Cortical visual areas in monkeys: location, topography, connections, columns, plasticity and cortical dynamics. Philosophical Transactions of the Royal Society B 2005, 709-731.

17] Sereno MI, Dale AM, Reppas JB, Kwong KK, Belliveau JW, Brady TJ, Rosen BR and Tootell RB Borders of multiple visual areas in humans revealed by functional magnetic resonance imaging. Science 1995, 268, 889-893.

18] Smith AT, Singh KD, Williams AL and Greenlee MW Estimating receptive field size from fMRI data in human striate and extrastriate visual cortex. Cerebral Cortex 2001, 11, 1182-1190.

19] Brewer AA, Liu J, Wade AR and Wandell BA Visual field maps and stimulus selectivity in human ventral occipital cortex. Nature Neuroscience 2005, 8, 1102-1109.

20] Larsson J and Heeger DJ Two retinotopic visual areas in human lateral occipital cortex. The Journal of Neuroscience 2006, 26, 13128-13142.

21] Goodale MA and Milner AD Sight Unseen OUP, Oxford 2004

2.0 UP WITH GEOMETRY AND DOWN WITH ALGEBRA IN ALL THINGS VISUAL

2.1 Algebra and geometry

In the early stages of the visual system the neurons are topographically organised, and part of the processing carried out by these stages involves forming representations of the spatial configurations in the retinal image which can be used by subsequent stages of the system. Elementary spatial configurations can be described quantitatively in two complementary ways. The geometric approach takes individual points and describes how they are joined together to form lines and how lines are joined together to form surfaces. The algebraic approach takes the equations of pairs of surfaces and uses their intersections to form lines and uses the intersections of three surfaces to specify points. The way in which the visual system handles its input at successive stages along its pathway will depend on the way in which three dimensional structures are represented by it and attempting to distinguish between the use of the geometric and algebraic forms of representation is a good starting point.

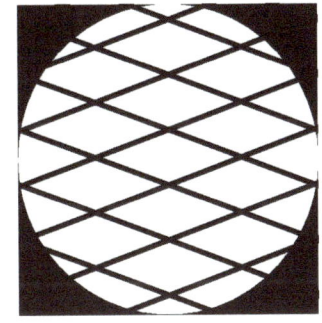

To begin with the geometric approach. A pair of numbers (x_1, x_2) can be used as the coordinates of a point in 2 dimensional space which has a position specified by a movement of an amount x_1 along the horizontal axis followed by an amount x_2 along the direction of the vertical axis, with the order in which the movements are made being unimportant. Similarly, a triplet of numbers (x_1, x_2, x_3) can be used to specify the coordinates of a point in everyday three-dimensional space in terms of movements along the horizontal, vertical and depth directions. Any displacement of a point is an example of a vector, which has both a size and a distinct quality, in this case direction. Over and above these properties, vectors can be added together to form another vector. In the text, vectors will be denoted by bold typeface. Using the vector concept, a point P in the plane which is specified by a pair of coordinates (x_1, x_2) can also be specified by a vector **p** which describes the change of position from the origin to the point by a sum of multiples of vectors in the horizontal (**h**) and vertical (**v**) directions, as illustrated in figure 2.1.

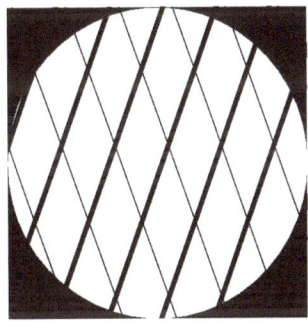

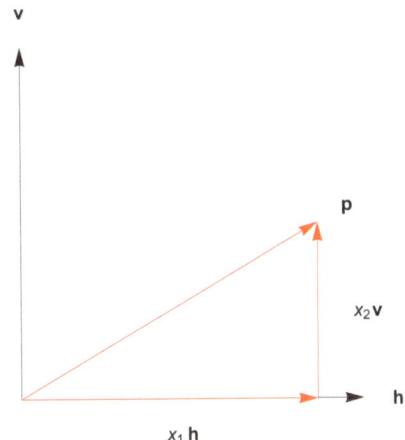

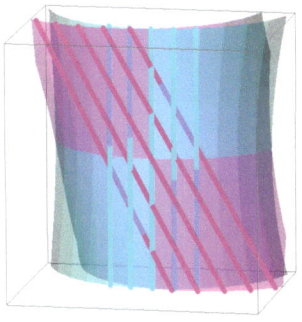

Figure 2.1 Diagram of how any point in the plane can be specified by a sum of horizontal and vertical displacements.

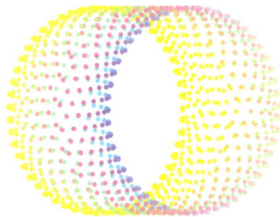

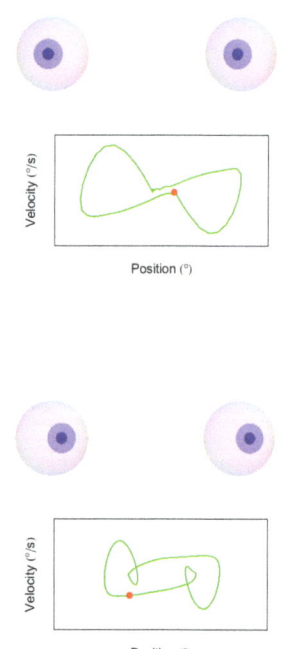

In this case, applying one vector after another corresponds to vector addition and the vector p can be specified by the vector equation:

$$\mathbf{p} = x_1\mathbf{h} + x_2\mathbf{v}$$

The sizes of the horizontal and vertical vectors are chosen to be one unit of length, so that the coordinates specify the displacements along the axes in multiples of units of length. In a similar fashion, a vector representation of any point in three dimensional space can be generated by adding a third depth vector to the set of horizontal and vertical vectors, which is perpendicular to both the other two vectors and also has unit length. The collection of points on a line can be specified by adding multiples of a vector **b** in the direction of the line to a vector **a** to a point on the line, as shown in figure 2.2. Using this approach, any point on the line can be specified by the vector equation:

$$\mathbf{p} = \mathbf{a} + x\mathbf{b}$$

where x is a parameter which specifies where on the line the point is. Similarly, any point on a plane surface in three-dimensional space can be specified by a sum of multiples of two vectors **b** and **c** which point in different directions on the surface.

$$\mathbf{p} = \mathbf{a} + x\mathbf{b} + y\mathbf{c}$$

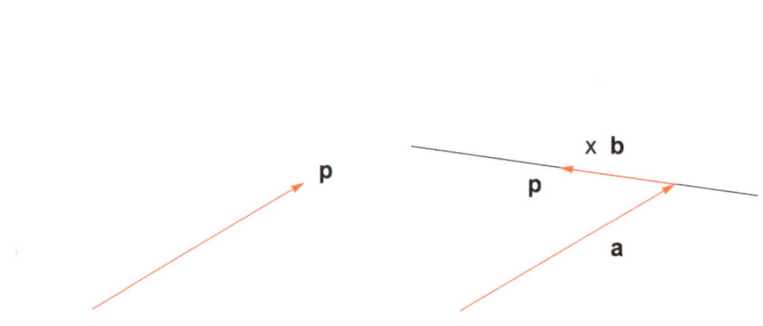

Figure 2.2 Vector descriptions of a point and a straight line on a plane surface.

The vector concept is not limited just to geometrical configurations but can be applied to any qualitatively different measurements which can be added together. In a typical colour matching experiment the subject views a small stimulus which is usually restricted to around 2 degrees in diameter so that only the fovea is stimulated. The field is split into two halves and a pair of lights A and B, presented in one half, have to be matched in colour to a pair of lights C and D, presented in the other half. The main finding of colour matching experiments, referred to as the trichromatic law of colour matching, is that the colour of any stimulus can be matched by an additive mixture of three fixed primary colours [1]. The experimental findings can be described geometrically, by treating each colour stimulus as a vector, with the direction of the vector representing the qualitative nature of the colour and the size of the vector representing the amount of the colour. The vectors $\mathbf{u}_1$, $\mathbf{u}_2$ and $\mathbf{u}_3$ are given by the functions of wavelengths $\mathbf{u}_1[\lambda]$, $\mathbf{u}_2[\lambda]$, and $\mathbf{u}_3[\lambda]$ respectively, which describe the spectral power distribution of the fixed primary light sources. In vector notation, the trichromatic law states that any colour **c** can be matched by a sum of multiples of $\mathbf{u}_1$, $\mathbf{u}_2$ and $\mathbf{u}_3$:

$$\mathbf{c} = a_1\mathbf{u}_1 + a_2\mathbf{u}_2 + a_3\mathbf{u}_3$$

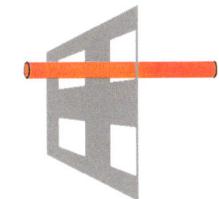

If one of the coordinates a_1, a_2 or a_3 is negative, then one can move the term involved to the other side of the equation, to obtain a description of the physical situation where one pair of colours is matched by another pair.

In 1931 a standard colorimetric observer was defined by the 'Commision Internationale de l'Eclairage' (abbreviated to C.I.E). The experimental results from many colour matching experiments were combined by expressing them all in terms of colour matches with three monochromatic light sources **r**, **g** and **b**, located at wavelengths of 700, 546 and 435.8 nanometres respectively. The relative amounts of the sources were set so that a colour mixture made by equal amounts of r, g and b appears white. The coordinates $\mathbf{r}(\lambda)$, $\mathbf{b}(\lambda)$ and $\mathbf{g}(\lambda)$ required to match a light with a unit radiant power at wavelength λ are functions of λ referred to as colour matching functions. The colour matching functions which form the specification of the C.I.E. observer are plotted in figure 2.3.

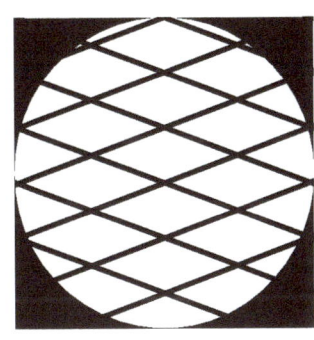

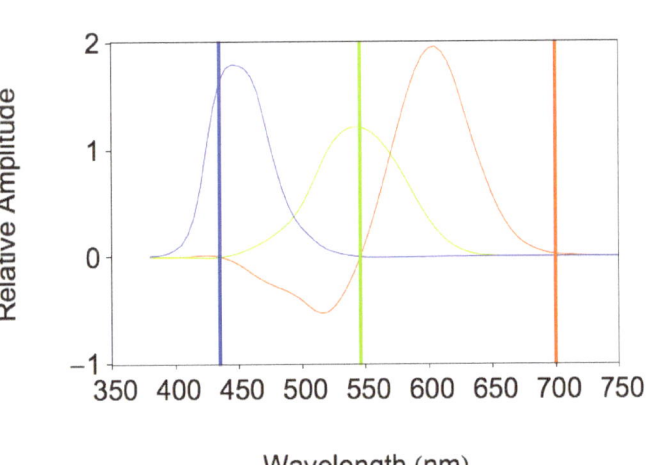

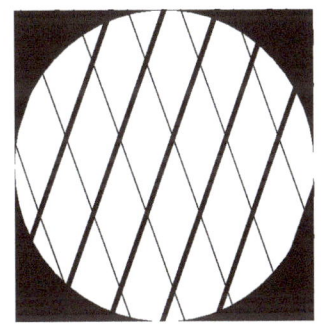

Figure 2.3. Colour matching functions of the C.I.E. observer.

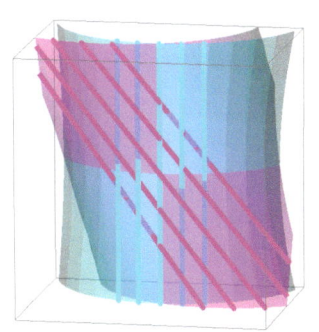

Moving away from the idea of a vector as just a displacement in two or three dimensions allows one to define an n-dimensional space as consisting of all the points which can be specified by a sum of multiples of n fixed vectors $\mathbf{v}_1$, $\mathbf{v}_2$ … $\mathbf{v}_n$, which are referred to as basis vectors. A restriction on this definition is that each of the basis vectors are qualitatively different, so that no basis vector is equal to a sum of multiples of the other basis vectors. The simplest example of an n dimensional space and is given by ordered sets of n numbers $(x_1, x_2 … x_n)$. Every vector in this space can be written as a sum of multiples of the n basis vectors $(1, 0, … 0)$, $(0,1, … 0)$, … $(0, 0, … 1)$.

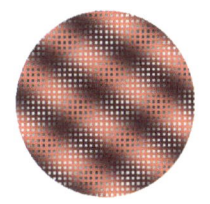

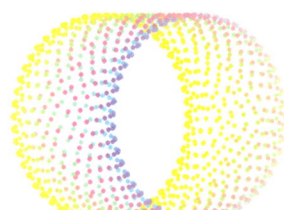

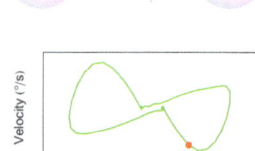

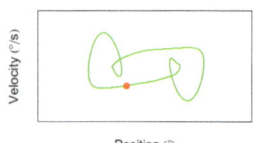

To return to the specification of spatial configurations. In the algebraic approach the coordinates of an n-dimensional vector **a** can be used to specify a linear equation of the form:

$$a_1x_1 + a_2x_2 + \ldots + a_nx_n = y$$

where $(x_1, x_2, \ldots x_n)$ are unknown and y is a constant. The implications of the equation are easiest to visualise in a two dimensional space. The points which satisfy the equation lie along a straight line which is perpendicular to the vector with coordinates (a_1, a_2) and a distance along the direction of the vector determined by y. So any straight line can be described by one equation. As illustrated in figure 2.4, any point can be specified by a pair of equations which describe a pair of straight lines that intersect at the point. With three dimensions, each linear equation specifies a plane surface. In n-dimensional space the surface of possible solutions to the linear equation is referred to as a hyperplane.

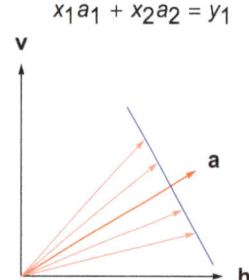
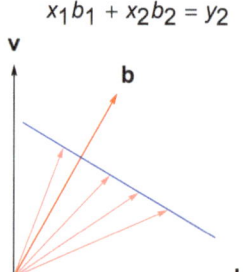
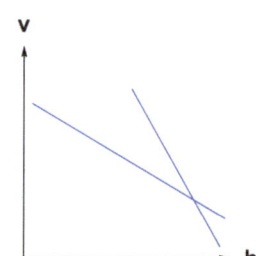

Figure 2.4. Algebraic coding of the position of a point on a plane. Two vectors **a** and **b** are shown in red. Each vector is used to specify a linear equation. The possible solutions to each equation lie along a line which is perpendicular to the vector and are plotted in blue. The coordinates of the point at the intersection of the two lines corresponds to the solution of the equations.

So does the visual system use geometric or algebraic representations of solid objects? Answering this question is not straightforward because the neural mechanism is sufficiently flexible to be capable of using either form of representation. A simple model of a neuron has a number of inputs, which describe the effects of the synapses on the dendrites and cell body of the neuron, and a single output as shown in figure 5. The individual inputs are weighted to represent the differing synaptic strengths of the inputs and the sum of these weighted inputs corresponds to the potential across the membrane of the cell body. Let the levels of the n inputs be denoted by $x_1, x_2 \ldots x_n$ and the levels of the output be denoted by y. Further, let the weights associated with each of the inputs be denoted by $w_1, w_2 \ldots w_n$ and assume that the output depends on the sum of the inputs, then:

$$y = w_1x_1 + w_2x_2 + \ldots + w_nx_n = \mathbf{w}.\mathbf{x}$$

The set of weights $w_1, w_2 \ldots w_n$ define a vector **w** in the n-dimensional space of the inputs and the response of the neuron depends on the amount of the input vector **x** in the direction of the weight vector, a quantity referred to as the projection of the input vector onto the weight vector.

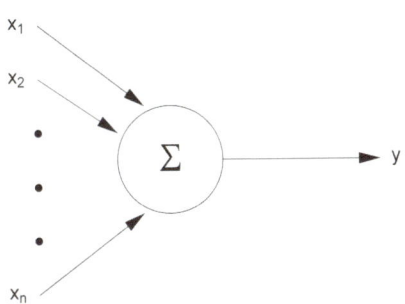

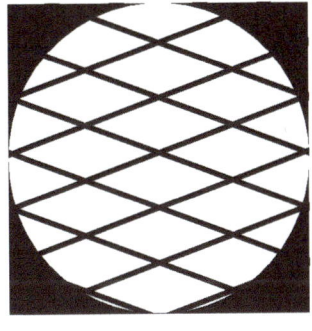

Figure 2.5 Components of a simplified model of a neuron. Each input x_i is multiplied by an associated weighting factor w and the output is equal to the sum of the weighted inputs. This quantity is referred to as the inner product and is denoted by **w.x**. This numerical quantity is proportional to the cosine of the angle between the two vectors. It is zero when the vectors are perpendicular and greatest when they are parallel.

Given the firing rate y of a neuron, the possible input values x1, x2 ... x_n which could have resulted in the output y are all solutions to the equation:

$$y = \mathbf{w.x}$$

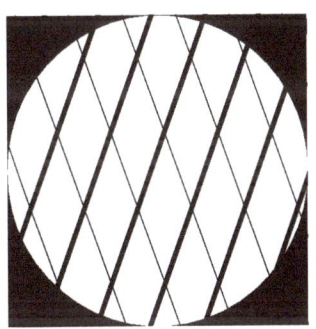

So an algebraic representation of a point on a flat surface (x_1, x_2) can be found by solving the pair of equations which describe the outputs (y_1, y_2) of a pair of neurons with weight vectors (a_1, a_2) and (b_1, b_2), provided that the weight vectors have different directions, as illustrated in figure 2.4.

The vector coding of a point on a plane can be implemented neurally by summing the projections onto the weight vectors of a pair of neurons with preferred orientations which are perpendicular to each other so they have no projection onto each other. The projection of the input vector **x** onto the weight vector $\mathbf{w}_1$ of one neuron gives one of the coordinates of the point. Adding a second neuron with a weight vector $\mathbf{w}_2$ which is perpendicular to the weight vector of the first neuron gives the second coordinate. This implementation is a neural version of the procedure shown in figure 2.1.

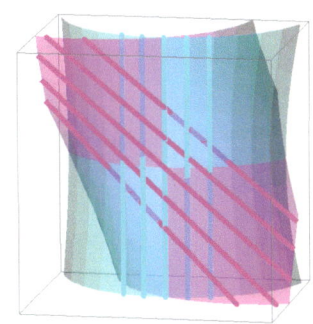

2.2 Motion and stereoscopic parallax

Another approach to investigating how the visual system represents spatial configurations is to consider the differences in the views of a scene which result from changes in the position of an observer. The most common form of parallax occurs when the fovea is kept on a particular part of the scene as the position of the eye is changed, as illustrated in figure 2.6. The parallax illustrated in this figure could arise from monocular viewing of an object as one moves past it, in which case the parallax is referred to as motion parallax, or from binocular viewing of an object, in which case the parallax is referred to as stereoscopic parallax.

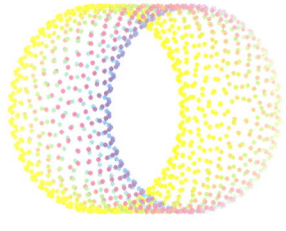

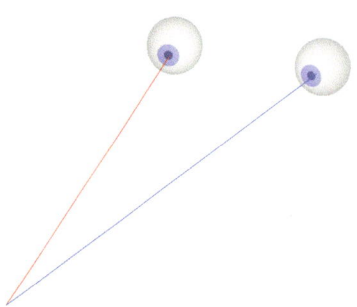

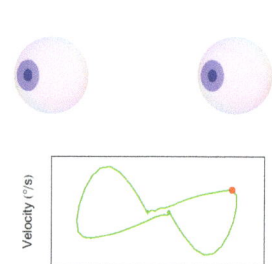

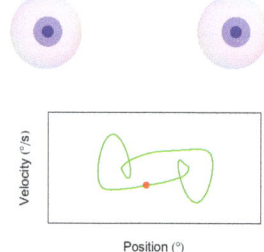

Figure 2.6 A diagram of the geometry of parallax which occurs when an eye views a point of fixation from two different positions.

The key to interpreting the different views of a scene lies in identifying features which are not viewpoint dependent [2]. For instance, if one sees a straight line from a particular viewpoint, then it could be a projection of a line which is bent in depth, but this will become obvious if one moves one's head from side to side. Similarly two lines joining at a junction could have come from two 3D lines which happen to have ends which project to the same point in the particular viewpoint. Again this will become obvious with a change in viewpoint. Features which are independent of the viewpoint can be reliably used to infer the nature of the objects in the scene. It is almost always true that a straight line in a picture is the projection of a straight line in the scene, and that a pair of lines which meet at a point in the picture are the projection of a pair of lines which meet at a point in the scene. Although there are many other such properties, such as collinear lines in the picture being projections of collinear lines in the scene, the straight line and junction properties are the most relevant to the visual system, because they can be identified by local operations by individual cells. For the local processing of parallax information, straight lines and line terminations are the features which have to be given a neural representation.

A pair of lines which meet at a junction are usually seen as perpendicular to each other in three dimensions. This is because there are two orientations of a plane through the lines for which the changes in the positions of the lines are least with changes in the position of the head. In the case of the folded card these two orientations correspond to the convex and concave appearances of the card. Another instance of the striking tendency to see co-terminating edges at right angles to each other is provided by the Ames window which is illustrated in the first flip book. The Ames window demonstration consists of a trapezoidal-shaped piece of card with a window drawn on it [3]. The card is rotated around a vertical axis at 2 revolutions per minute. The window appears to oscillate rather than rotate, being assumed to be rectangular. When a biro is placed through one end of the planes of the window, it appears to bend, elongate and partly wrap itself around the window frame.

When dealing with parallax images, the visual system shows it has some ideas of its own and appears to switch between the vector and algebraic geometry representations according to its own whims. For instance an infinitely long straight line could be moving in any direction and yet give rise to the same sequence of images, because the component of movement along the line does not change the image, as illustrated in figure 2.7. Despite this ambiguity, usually one sees a straight line moving perpendicular to its orientation. The ambiguity is referred to as the aperture problem, because it is arises whenever the line terminations are hidden. It can be resolved if the terminations of the lines are visible or if there are lines with more than one orientation.

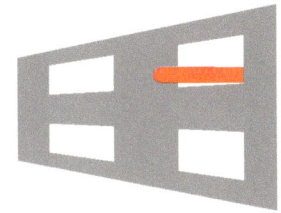

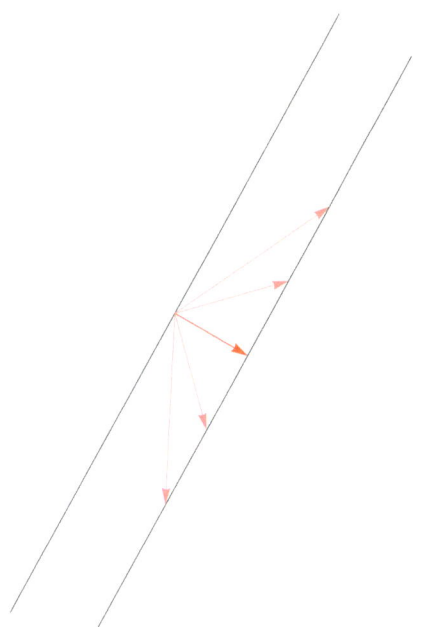

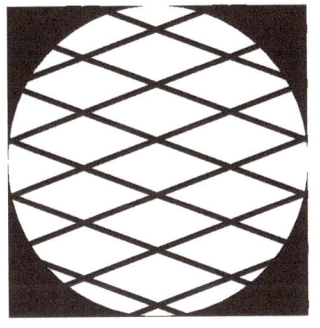

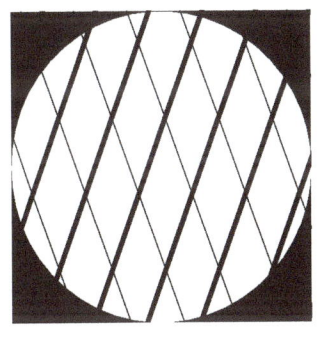

Figure 2.7 Illustration of some of the possible directions of movement of a straight line which are indistinguishable because movements of the line along its length make no difference to the stimulus. The line usually appears to move perpendicular to its orientation[4].

A single moving line can be represented by an extension of the vector coding scheme. The response of neurons in cortical area V1 depends on the projection of the motion onto the preferred movement direction of the cell. A further biological constraint on coding is that neurons always have a positive firing rate so that for every neuron with a weight vector **w** another neuron with direction −**w** is required to represent the entire range of a linear neuron. The response of the i th neuron can be described by the equation:

$$y^+{}_i = \mathbf{x}.\mathbf{p}_i$$

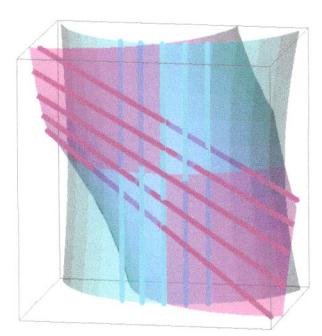

Where **x** is the vector describing the velocity of the line, and y^+ is a nonlinear function that only gives a non-zero value when $\mathbf{x}.\mathbf{p}_i$ is positive. With this response function, the tuning of a neuron can be plotted on a polar diagram as shown in figure 2.8.

Because neurons in V1 are tuned to the entire range of directions of movement many neurons respond to a single stimulus. If one imagines this population of k neurons to be made up of pairs of neurons with perpendicular weight vectors then it follows that each pair of perpendicular neurons will give the correct vector coding of the point. The population vector **ñ** is defined as the vector sum of the weight

vectors of the neurons multiplied by their outputs [5]:

$$\tilde{\mathbf{n}} = y_1\mathbf{w}_1 + y_2\mathbf{w}_2 + \ldots + y_k\mathbf{w}_k$$

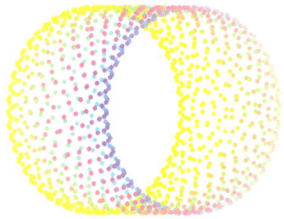

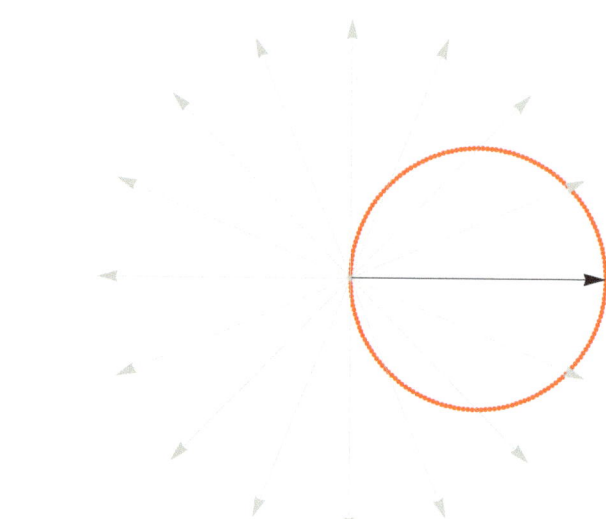

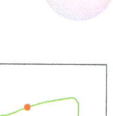

Figure 2.8 Polar diagram of the cosine vector tuning similar to that of velocity sensitive neurons in MT. The grey arrows show the preferred velocity directions of a population of 16 neurons. The red circle is a polar plot of the response of the neuron with the preferred direction shown in black. The response drops off as a cosine function of direction which generates a circular plot in polar coordinates.

This scheme predicts that whenever there are lines with more than one orientation present movement in a single population vector direction will be seen. This phenomenon is illustrated in flip book 2. But the appearance of a single direction of motion could also be taken as evidence for algebraic coding of the movement. Each set of moving lines of a particular orientation can be used to define a single linear equation, the solutions of which correspond to the possible velocity with which the lines are moving, and a pair of sets of moving lines will have a unique solution [6]. A problem with this interpretation is that if the lines have orientations close to the direction of movement then they appear to slide over each other rather than move together. This is illustrated in flip book 3.

The motion parallax examples demonstrate that the vector and algebraic coding frameworks capture different aspects of the underlying neural mechanism. Whilst the population vector will give the correct estimate of the direction of the input vector, it will usually overestimate the length of the input vector because it is based on the sum of many pairs of neurons. One way round this drawback is to divide each of the outputs by the sum of the lengths of the weight vectors [7]. The population code then represents the velocity of the stimulus.

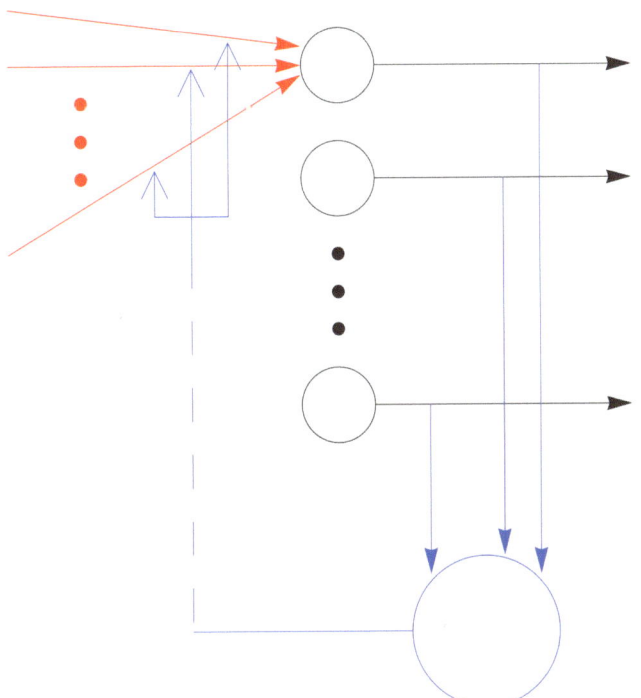
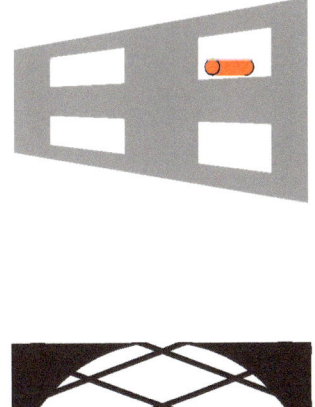
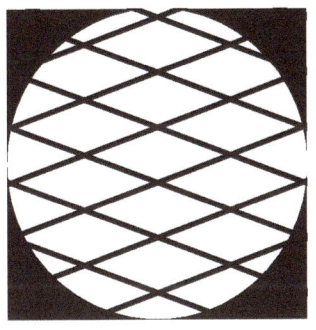

Figure 2.9 A circuit which could explain the responses of neurons in MT. The responses of the neurons are pooled and used to divide the inputs from movement sensitive neurons in V1 to obtain a population vector code.

A similar combination of aspects of vector and algebraic representations is found with stereoscopic parallax. The simplest examples of stereoscopic views of straight lines consist of the cases where the images of the straight lines appear in identical positions in each eye. A set of such lines, all with the same retinal orientation will lie on a quadric surface in space. With the eyes symmetrically converged to a point on the midline in the horizontal plane, the surface ruled by lines which have vertical retinal images is a cylinder. Figure 2.10 shows some examples of the lines lying on this surface. For an oblique retinal orientation the surface ruled by the lines is an hyperboloid of one sheet. The changes in 3D shape with changes in retinal orientation are illustrated in flip book 4. For horizontal retinal lines the surface degenerates into two planes, one horizontal and one vertical. All these surfaces intersect in a curve known as the horopter, which consists of a circle through the point of fixation and a vertical line through the point of fixation [8].

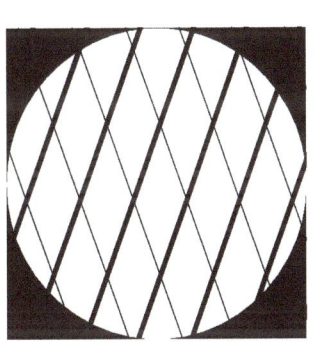

Next consider what happens when one of the sets of parallel retinal lines is displaced nasally. In this case, the only possible stationary object which could give rise to this pattern of retinal images is a set of three dimensional lines which pass through a circle lying inside the horopter circle. Similarly, if one set is displaced temporally, then the object which gives rise to the retinal images consists of a set of three dimensional lines which pass through a circle lying outside the horopter circle. So in the case of stereoscopic parallax, the position of straight lines in space are completely determined by the horizontal differences between the positions of the images in the left and right eyes.

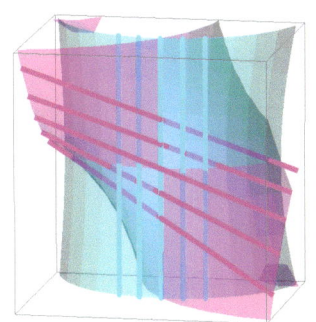

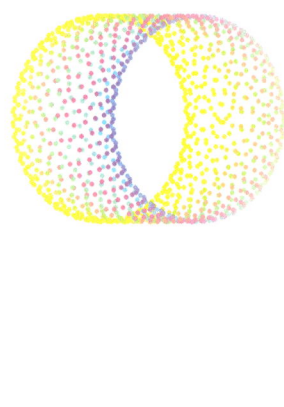

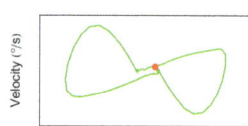

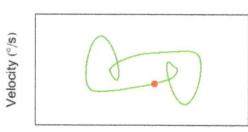

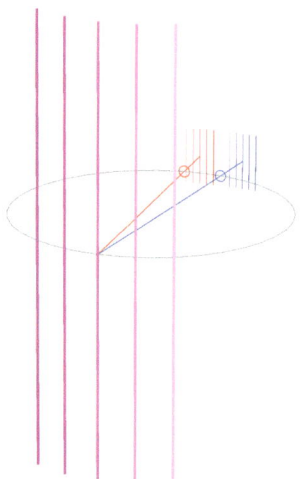

Figure 2.10 Examples of the vertical lines which cast images on identical positions on each retina. The entrance pupils of the left and right eyes are represented by blue and red circles respectively. The line through the centre of the fovea and the centre of the entrance pupil of the left eye intersects with the corresponding line for the right eye at the point which is being looked at. The grey line marks the circular portion of the horopter curve.

However this does not mean that one can immediately tell the positions of straight lines in space from a pair retinal images. For example, hammer two nails into a straight piece of wood approximately 15 millimetres apart at a distance of 50 centimetres from one end and then view the nails with the piece of wood touching one's nose and pointing straight ahead, as shown in figure 2.11. Provided the nails are aligned so that they appear at the same level, one sees the 'ghost' images of the nails in which the image of the further nail in one eye is matched with the image of the nearer nail in the other eye [9]. In order to decide which scene a pair of stereoscopic images could have arisen from, the brain has to match each part of the image in the left eye with the part of the image in the right eye which comes from the same bit of the scene. This task is referred to as the correspondence problem [10].

The difference between the positions of the images of an object in the left and right eyes is referred to as the disparity of the image. One way in which the brain simplifies the matching task is by only allowing matches in which the disparities are small. If the disparities are large the retinal images in the left and right eyes are seen as coming from two different objects. If two pencils are held upright and 100 millimetres apart along the straight ahead direction, then two types of double vision can be distinguished. If the nearer pencil is fixated then the further pencil appears as a double image. Closing one's left eye demonstrates that the image from the right eye appears on the right, hence these images are referred to as uncrossed double images. If the further pencil is fixated then the nearer pencil appears as a double image. In this case, closing one's left eye demonstrates that the image from the right eye appears on the left, hence these images are referred to as crossed double images.

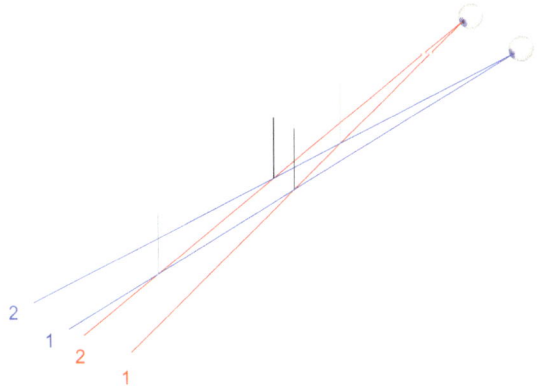

Figure 2.11 The double nail illusion. Two different configurations of vertical rods which give rise to the same stereoscopic parallax. If the visual directions are numbered from left to right, then the two configurations involve matching directions 1 and 2 in the left eye with directions 1 and 2 (black lines) and 2 and 1 (grey lines) in the right eye.

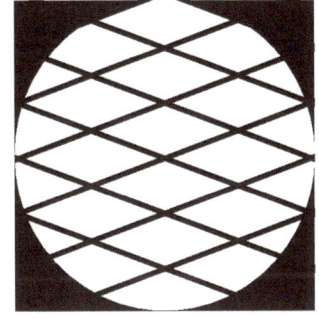

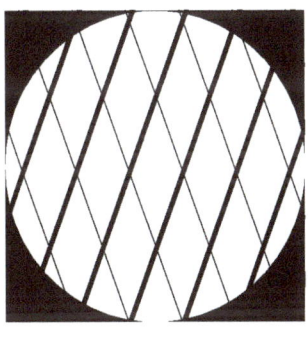

Once the correspondence problem has been solved, the three dimensional position of each point of a spatial configuration could be found algebraically by finding the intersection of the rays through each pair of image points. The alternative vector coding approach would involve computing the responses of neurons tuned to specific disparities and then using divisive inhibition to eliminate the false matches.

2.3 Oculomotor coding

The rapid eye movements which are used to transfer one's gaze from one object to another are referred to as saccades. The most direct pathway for the control of saccades runs from the retina to the superior colliculus, then on to the brainstem, and finally on to the oculomotor nuclei. The superior colliculus is a layered structure with visual cells located superficially and motor cells located in intermediate and deeper layers. The receptive fields of cells in the sensory layer are large and respond to stationary spots or spots moving in any direction in the visual field. Cells in the deep layer respond prior to a saccade in a particular direction and with a particular amplitude. The range of amplitudes and directions of saccades with which the firing of a cell is associated is independent of the initial direction of the line of fixation, and is referred to as the movement field of the cell. Both these layers are topographically organised and the maps are in register.

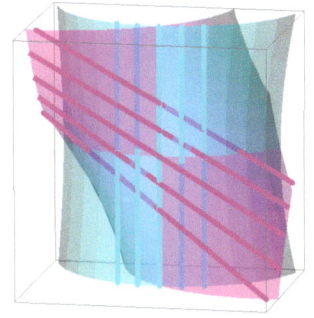

Stimulation of the deep layer of the superior colliculus with pulse waveforms results in a change in the direction in which the eye is looking towards the contralateral visual field. The saccadic eye movements are of a fixed amplitude and direction, which depend only on the site of the stimulating electrode within the colliculus. A long pulse train leads to a staircase of saccades, all of the same amplitude and direction. When two sites are stimulated simultaneously, typically a single saccade occurs which has an amplitude and direction which is a weighted mean of the saccades elicited by stimulation of each site on its own, the weights being in proportion to the relative intensity of the stimulation of the two sites [11]. Since stimulation of the deep layer of the superior colliculus results in a saccade to a

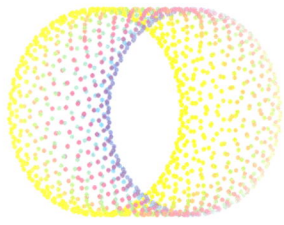

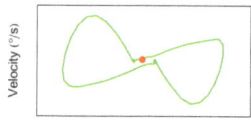

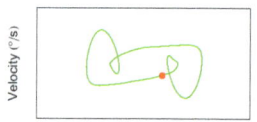

position in the visual field which is approximately the same as that of the receptive fields in the overlying sensory layer, the evoked saccades can be used to recover the topography of the afferent mapping [12]. The overall saccade vector is given by the vector sum over all the cells in the colliculus of the saccade vector associated with each cell, weighted by a gaussian function centred at the target location, as illustrated in figure 2.12 . Integration of the preferred direction of each cell over the duration of the saccade gives the projection of the preferred direction onto the direction of the target displacement. In this case the overall inhibition depends on the sum over the duration of the movement of the output of the colliculus, which ensures that the same number of spikes are produced even when the saccades is perturbed by blinks [13,14].

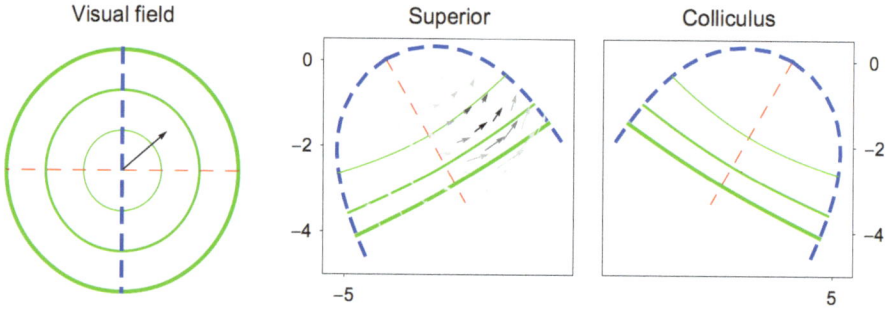

Figure 2.12 Coding of the amplitude and direction of a saccadic eye movement in the superior colliculus.

2.4 References

1] Wright WD The Measurement of Colour Adam Hilger, London, 1944

2] Hoffman DD Visual Intelligence. W.W.Norton & Company, Inc., New York, 1998.

3]] Ittelson WH The Ames Demonstrations in Perception (originally published 1952) together with an Interpretive manual by A.Ames. Jr. (originally published 1955). Hafner Publishing Company, New York, 1968.

4] Wallach H Uber visuell wahrgenommene Bewegungsrichtung. Psychologische Forschung 1935, 20, 325-380.

5] Schwartz AB, Kettner RE and Georgopoulos AP Primate motor cortex and free arm movements to visual targets in three-dimensional space. I. Relations between single cell discharge and direction of movement. Journal of Neuroscience 1997, 77, 826-852.

6] Adelson EH and Movshon JA Phenomenal coherence of moving visual patterns Nature, 1982, 300, 523-525.

7] Heeger DJ, Simoncelli EP and Movshon JA Computational models of cortical visual processing. Proceedings of the National Academy of Science of the U.S.A 1996, 9, 623-627.

8] Helmholtz H von Treatise on Physiological Optics. 1910. Translated by J.P.C.Southall, New York, Dover, 1924.

9] Krol JD and Grind WA van de The double-nail illusion: experiments on binocular vision with nails, needles, and pins. Perception 1980, 9, 651-669.

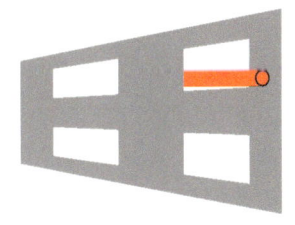

10] Julesz B Foundations of Cyclopean Perception. University of Chicago Press, Chicago, 1971.

11] Robinson DA Eye movements evoked by collicular stimulation in the alert monkey. Vision Research 1972, 12, 1795-1808.

12] Ottes F, van Gisbergen JA and Eggermont J Visuomotor fields of the superior colliculus: a quantitative model. Vision Research 1986, 26, 857-873.

13] Tabareau N, Bennequin D, Berthoz A, Slotine J-J and Girard B Geometry of the superior colliculus mapping and efficient oculomotor computation. Biological Cybernetics 2007, 97, 279-292.

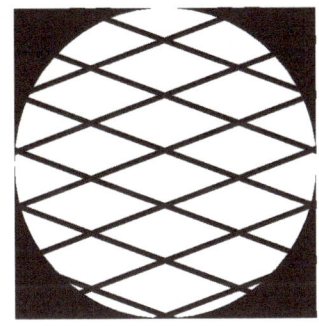

14] Goossens HHLM and Van Opstal AJ Dynamic ensemble coding of saccades in the monkey superior colliculus. J. Neurophysiol. 2006, 95, 2326-2341.

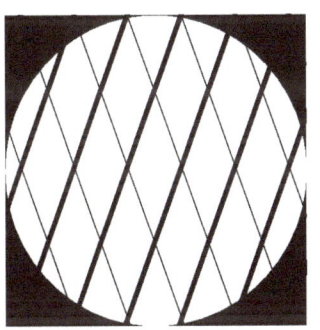

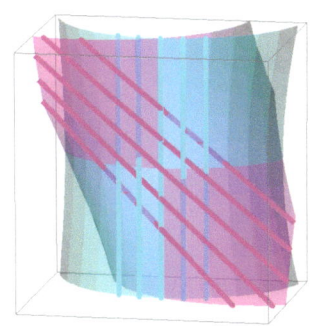

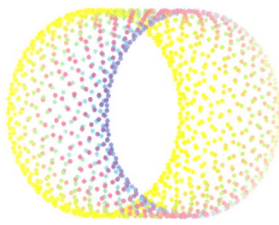

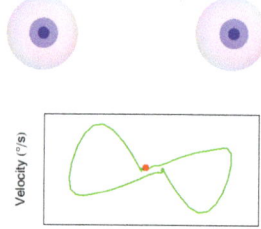

-
-
-
-
-
-

3.0 SEEING THINGS

3.1 Reconstruction, reconstruction, reconstruction

A distinctive feature of visual neurons is that each neuron receives inputs from a particular area of the retina, referred to as its receptive field [1]. In the early stages of the visual pathway the strengths of the connections associated with each position in the receptive field can be found experimentally by flashing small spots of light at each position and measuring the response of the neuron. In the retina the receptive fields of the cells have a concentric organisation consisting of a centre and a surrounding ring. The centre responds to a stimulus being turned either on or off, while the surrounding ring responds to a light action that is opposite to that of the centre. The output of the neuron depends on the balance between the stimulation of the centre and surround, for if a large light spot covers both the centre and surround then the response is less than with stimulation of either the centre or surround alone [2].

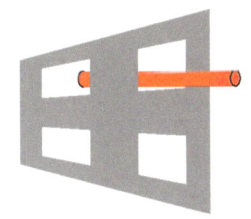

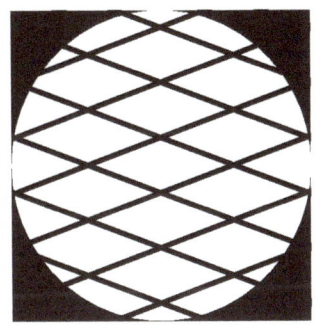

Because the retinal ganglion cells do not respond to uniform patches of illumination the information contained in the firing pattern of the cells is not the same as that contained in the optical image in the retina. This difference can lead to illusions in which the image appears different from the physical stimulus. The Hermann grid phenomenon, which was first reported in 1870 [3], is an example of such an illusion. The Herman grid consists of a regular array of black squares separated by white lines, as shown in figure 3.1. People usually report seeing illusory grey spots at the intersections of the white lines, although not at the intersection being looked at. An explanation of the appearance the Hermann grid in terms of the properties of receptive fields runs as follows. If the luminance across the receptive field is uniform, then the neurons will produce little or no response as centre and surround will both be stimulated. So the response of the neurons will be least for the insides of the black squares. The neurons will give maximal response when a light bar just covers the centre, because then the excitatory input will be greatest, and the inhibition less than when the receptive field centre just covers one of the intersections of the light bars. As the response of the neurons is less at the intersections, they appear darker than the bars [4].

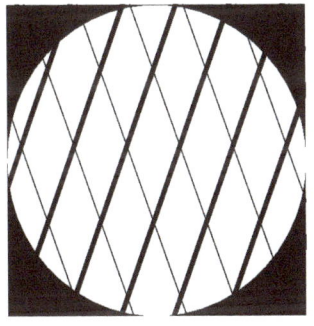

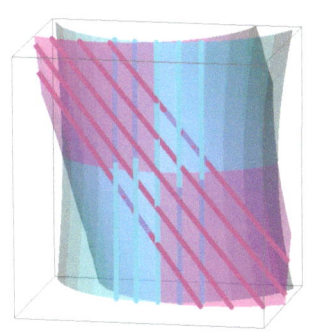

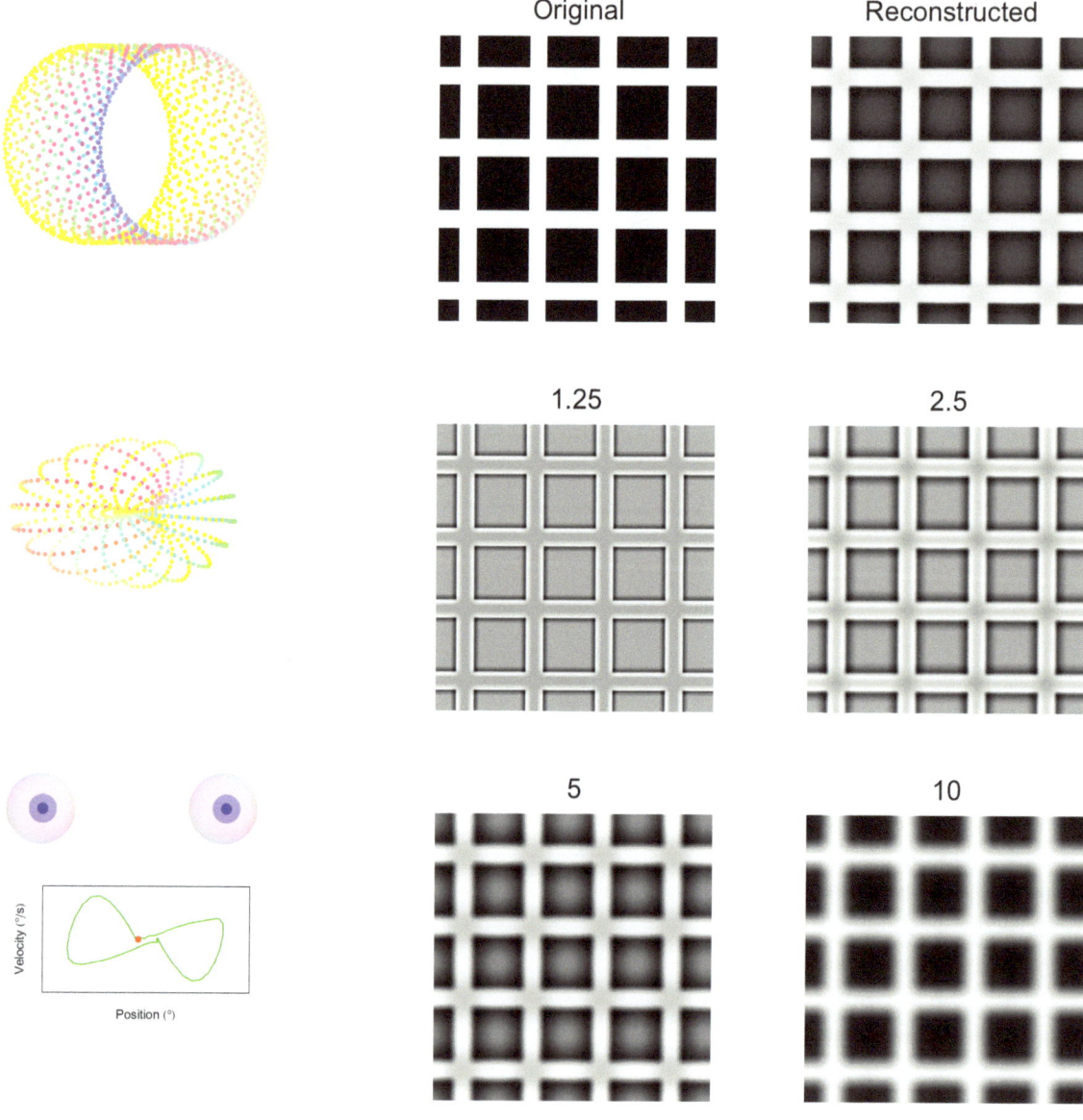

Figure 3.1 The Hermann grid illusion. The top row shows the original figure and the reconstructed figure made by adding together the responses of neurons with four different sized receptive fields. The figures below illustrate the responses to the pattern of each of the four different receptive field with centre radii of 1.25, 2.5, 5 and 10 minutes of arc.

The explanation of the Hermann grid in terms of receptive fields is very useful as it implies an experimental technique for measuring the dimensions of the receptive fields. Firstly, the grey spots should be most obvious when the bar widths exactly match the diameters of the receptive field centres. This occurs when the bars are 5 minutes wide. Secondly, the grey spots will become less visible if the inhibition by the surround drops off. So by decreasing the radial extent of the bars from the intersections, the diameter of the surround can be determined. The diameter is 20 minutes. These figures are not the dimensions of the receptive field of a single neuron, but rather the average of all of the receptive fields centred at a particular location. This average is referred to as a perceptive field and just as a receptive field can be used to estimate the response to different spatial patterns, the perceptive field can be used to estimate the appearance of different spatial patterns [5].

The centre-surround organisation of a perceptive field can be modelled by putting the weights near the centre of the perceptive field equal to a positive number and the weights further away to a smaller negative number, as plotted in figure 3.2. The extent of the centre and surround regions are obtained from the experimental findings with the Hermann grid and the relative sizes of the weightings of the centre and surround are balanced so that if the whole perceptive field is covered then the weighted sum is zero.

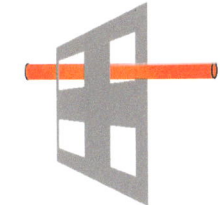

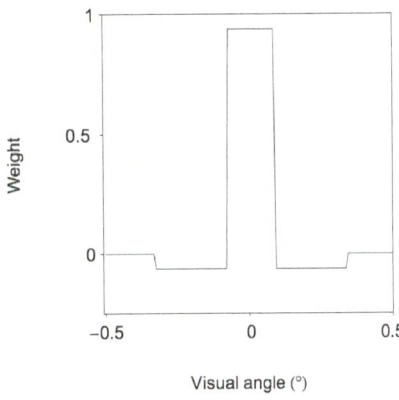
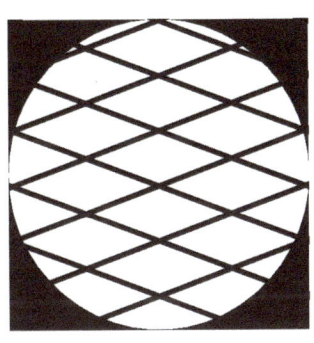

Figure 3.2 The weighting function of the human foveal perceptive field. The square on the left represents a patch of retina subtending a degree by degree visual angle. The positive weights in the centre are shown in blue and the negative weights of the surround are shown in reddish-brown. A cross-section of the weighting function taken at the level of the horizontal white line is shown on the right.

The responses of the model neurons, with a range of receptive field sizes, at every point of the Hermann grid is shown as a grey level plot in figure 3.1. Clearly, these responses have to be further transformed to recover the original appearance of the Hermann grid. But it turns out that simply adding the responses of set of neurons with a range of receptive field sizes gives a passable approximation to the original stimulus, as illustrated in figure 3.1, although in the case of the Hermann grid with the addition of the grey spots [6,7].

3.2 Informative features

The receptive fields of cortical cells are different from those of cells in the retina and lateral geniculate nucleus because they are elongated in one direction. A consequence of this receptive field organisation is that cortical cells typically do not respond best to spots of light but to bars and edges. Amongst cortical cells a distinction can be made between simple cells, which have receptive fields that can be explored with stationary spots of flashing light, and complex cells, which respond poorly to stationary spots of light but well to moving bars and edges [8]. Extending beyond the receptive field of a cell is a non-responsive region of the retina, which is typically 3 times the size of the receptive field. Although stimulation of the non-responsive region does not elicit a response from the neuron, it does alter the response of the neuron to a stimulus in its receptive field. The non-responsive region modulates the response of the cell by divisive inhibition, so that the response of the cell to a stimulus in its receptive field is reduced by an amount which depends on the level of stimulation of the non-responsive region [9,10].

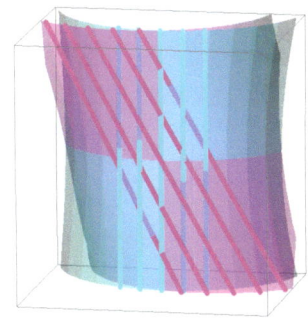

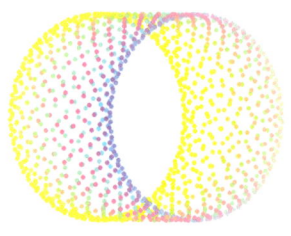

With elongated receptive fields it is possible to construct two alternative spatial weighting functions which both have the same tuning for spatial detail, as shown in figure 3.3. One is symmetric about the origin and the other is antisymmetric ($f(x) = -f(-x)$) about the origin. The symmetric weighting function gives maximum response to a narrow bar, whilst the antisymmetric weighting function will give maximum response to a step edge.

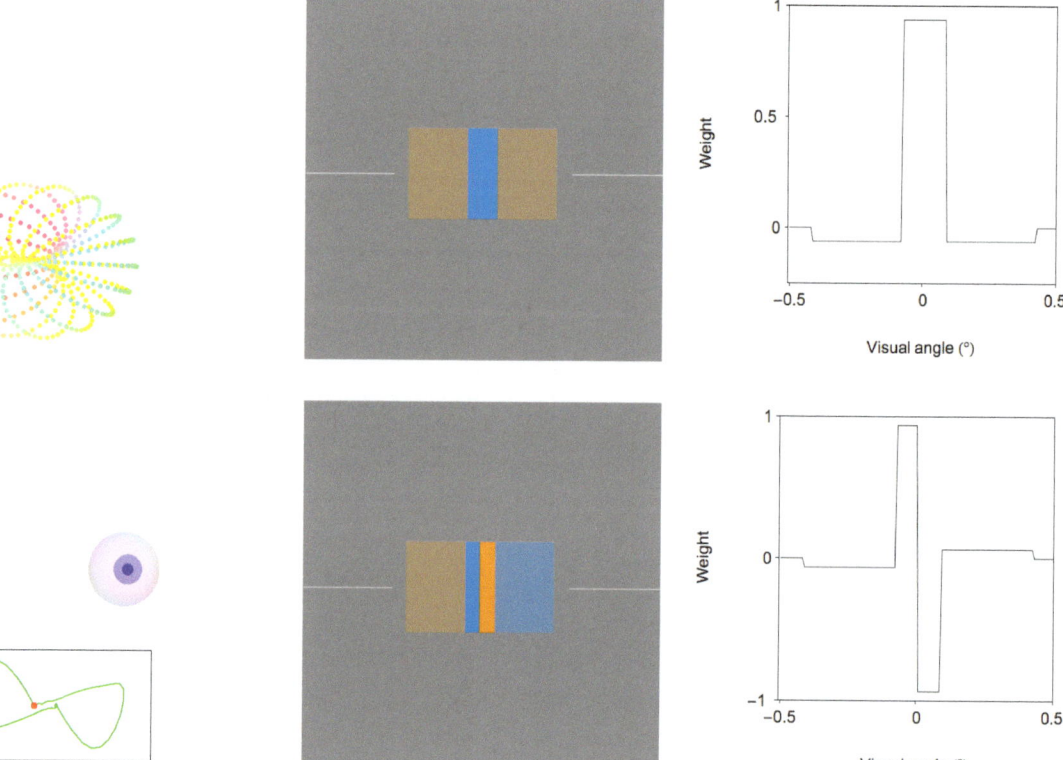

Figure 3.3 The weighting functions of symmetric and antisymmetric cortical receptive fields.

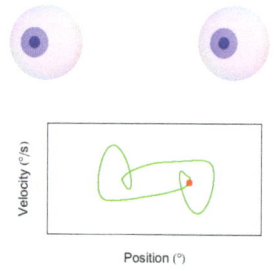

Visual perception depends on selective processing of the informative features of the retinal image. The cortical processing of the retinal image accentuates the informative features. An example of this accentuation process is given in figure 3.4. The original picture of the dog is shown at the top left. The output of a cortical network, with a cross-sectional centre-surround ratio the same as that of the foveal perceptive fields, emphasizes the eyes and nose. But you do not need any complicated neural modelling to recover the informative features [11]. Simply tracing over the features which are important for subsequent recognition of the dog give a drawing something like that shown on the lower right side of the figure.

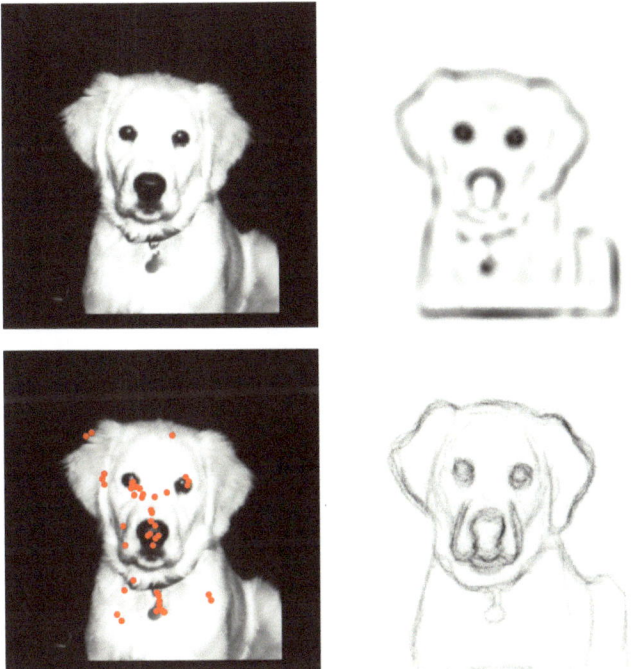
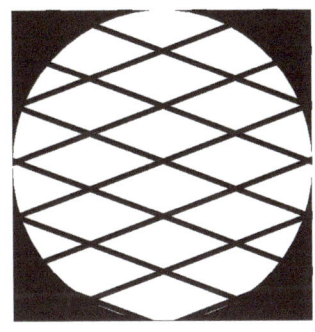

Figure 3.4 Selective processing of contours. The original picture is shown at the top left. The output of a cortical model is shown at the top right. The picture on the lower right is an average of a number of peoples' attempts at tracing over the picture of the dog. The final picture shows the points which were looked at by a subject during a period of 10 seconds.

The lateral connections in the cortex, which are responsible for the non-responsive regions, also give the network flexibility to switch between different patterns of acivity. With some stimuli the activity in the cortex can settle into more than one pattern. An example of such a stimulus is the Marroquin pattern which is produced by overlaying 3 square arrays of dots which are rotated at –60/0/+60 degrees away from the vertical [12]. This pattern is shown in figure 3.5. With free viewing of the pattern, subjects report seeing illusory circles appear and disappear. As well as the circles, patterns of squares and straight lines can be seen. In the output of the cortical model it can be seen that the circular features are arranged concentrically in the pattern and the locations at which the circles are seen are predictable from the model [13].

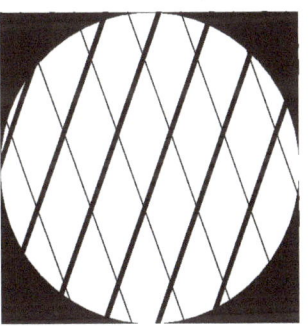
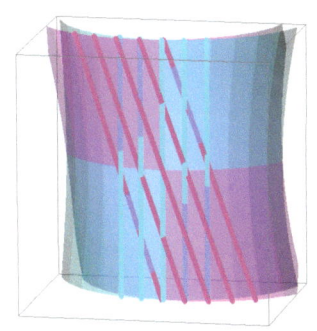

Figure 3.5 The Marroquin pattern and a possible cortical representation.

3.3 Eye movements and working memory

The eye is incessantly making small, unnoticed movements which are necessary to keep the retinal image from becoming stationary on the retina. Neurons adapt to stationary images so that eventually the images fade away. These eye movements can be noticed by looking steadily at the red dot at the centre of the Hermann grid shown in figure 3.6 and then after about 20 seconds looking at the white dot at the centre of a black square. A negative afterimage of the white lines is seen, which does not remain stationary, because of the movements of the eyes [14]. Another example is a pattern that was generated by a mathematical model of the pattern of lines in Bridget Riley's picture 'Fall' [15]. In this case the movements of the eyes stimulate motion sensitive neurons which in turn signal motion perpendicular to the orientations of the lines. This gives the impression of horizontal bands of movement.

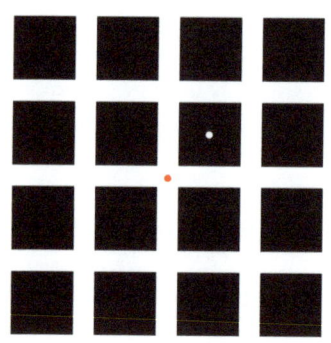

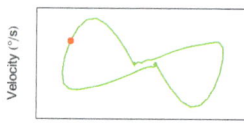

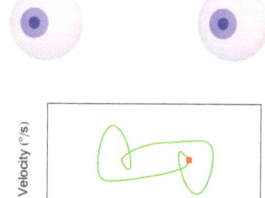

Figure 3.6. Perceptual demonstrations of the incessant movements of the eyes. See text for details.

The human eye has a fovea where visual resolution is greatest. As the fovea subtends a visual angle of around 5 degrees only a small portion of a scene can be seen in detail at any one time. Eye movements are made to centre the informative features of the scene on the fovea, and they can be used as an indicator of where attention is being directed [16]. The features that are fixated depend on the task which the subject is carrying out but, even with free viewing of pictures, attention is drawn to specific features. In the case of faces, the fixation eye movements are concentrated on the eyes, nose and mouth [17]. In more abstract images, the fixations are concentrated on the locations of the image with the most contours and the highest contrast [18]. Both of these tendencies are illustrated by the subject's eye movements shown in figure 3.4.

One way of tackling the question of how information about a scene is built up over successive glances is to consider what happens when one looks at silhouettes [19]. The information in silhouettes is contained in their contours and if it is assumed that the positions of all of the straight contour segments in the silhouette are known, then the task for the observer is to identify the orientations of all of the contours. The amount of information that the observer has about the orientation of an edge can be characterised by the probability distribution over all possible orientations of the edge. If the distribution is flat the observer has no information about the orientation of the edge, whereas if the distribution is sharply peaked then the observer has almost complete information about the orientation of the edge. According to this approach the next fixation position should be chosen to maximise the average information over all the edge points.

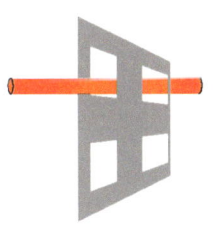

The information gathering approach can be applied to the Marroquin pattern by assuming that the pattern consists of circles at known locations and that each of the tangents at twelve locations on each of the circles have one of six possible orientations. Figure 3.7 gives an example of how uncertainty about the Marroquin stimulus is reduced with prolonged viewing. If the entire history of positions looked at contributes information then after thirty fixations there is little uncertainty left about the orientations of the tangents. But this does not tally with what one sees, which is circles appearing and disappearing at random. One possibility is that information from only a few of the last glances is held in memory [20]. If only information from the last six points looked at is used then the calculation of uncertainty suggests that first the subject becomes sure of the orientations of the circle at the centre of the pattern and then this certainty fades and the subject becomes sure about the orientations of a circle in the lower right quadrant of the pattern. This sort of behaviour would be consistent with that of a spatial array of neurons with locally excitatory lateral connections, which enable the formation of a patch of activity, and wider ranging inhibitory lateral connections, which stop the patch of activity spreading.

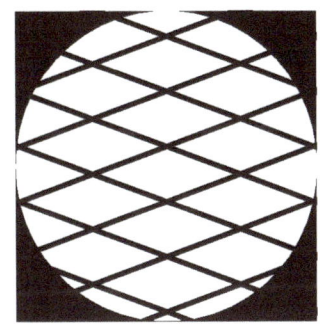

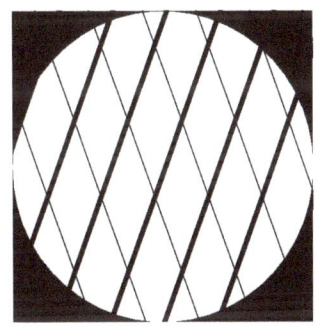

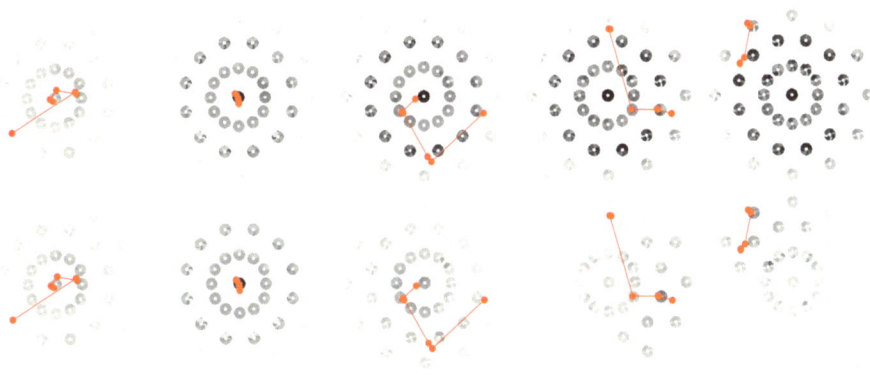

Figure 3.7 Illustration of the build-up of information about the orientations of the edges of the circles embedded in the Marroquin pattern. The uncertainty about the orientation of an edge is represented by a grayscale along which white is most uncertain and black is certain. Each picture in a row is separated from the previous figure by six fixations, which are plotted in red. In the top half of the figure all positions looked at contribute to reducing the uncertainty about the orientations. In the lower half of the figure only the last six positions looked at contribute.

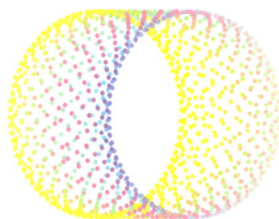

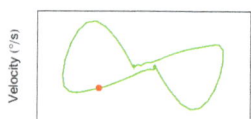

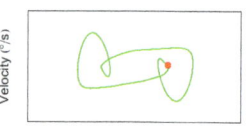

3.4 References

1] Hartline HK The response of single optic nerve fibers of the vertebrate vertebrate eye to illumination of the retina. American Journal of Physiology 1938, 121, 400-415.

2] Kuffler SW Discharge patterns and functional organization of mammalian retina. Journal of Neurophysiology 1953, 16, 37-68.

3] Hermann L Eine Erscheinung simultanen Contrastes. Pflügers Archiv für die gesamte Physiologie 1870, 3, 13-15.

4] Baumgartner G Indirekte Größenbestimmung der rezeptiven Felder der Retina beim Menschen mittels der Hermannschen Gittertäuschung Pflügers Archiv für die gesamte Physiologie 1960, 272, 21-22.

5] Spillman L Foveal perceptive fields in the human visual system measured with simultaneous contrast in grids and bars. Pflügers Archiv für die gesamte Physiologie 1971, 326, 281-299.

6] Blakeslee B and McCourt ME A multiscale spatial filtering account of the White effect, simultaneous brightness contrast and grating induction. Vision Research 1999, 39, 4361-4377.

7] Dakin SC and Bex PJ Natural image statistics mediate brightness 'filling in'. Proceedings of the Royal Society of London. Series B. 2003, 270, 2341-2348.

8] Hubel G and Wiesel TN Receptive fields, binocular interaction and functional architecture in the cat's visual cortex. J. Physiol. (London) 1962, 160, 106-154.

9] Sengpiel F, Baddeley RJ, Freeman TC, Harrad R and Blakemore C Different mechanisms underly three inhibitory phenomena in cat area 17. Vision Resarch 1998, 38, 2067-2080.

10] Cavanaugh JR, Bair W and Movshon JA Nature and interaction of signals from the receptive field center and surround in Macaque V1 neurons. Journal of Neurophysiology 2002, 88, 2530-2546.

11] Martin M, Fowlkes C, Tal D and Malik J A database of human segmented natural images and its application to evaluating segmentation algorithms and measuring ecological statistics. Computer Science Division (EECS) University of California, Berkeley. Report Number UCB/CSD-1-1133

[http://www.eecs.berkeley.edu/Pubs/TechRpts/2001/CSD-01-1133.pdf]

12] Marroquin JL Human visual perception of structure. Master's Thesis, Department of Electrical Engineering and Computer Science, MIT, 1976.

13] Wilson HR, Krupa B and Wilkinson F Dynamics of perceptual oscillations in form vision. Nature Neuroscience 2000, 3, 170-176.

14] Verheijen FJ A simple after image method demonstrating the involuntary multi-directional eye movements during fixation. Optica Acta 1961, 8, 309-311.

15] Zanker JM Looking at op art from a computational viewpoint. Spatial Vision 2004, 17, 75-94.

16] Mackworth NH and Morandi AJ (1967) The gaze selects informative details within pictures. Perception & Psychophysics 1967, 2, 547-552.

17] Yarbus, A.L. Eye Movements and Vision. (English translation by L.A. Riggs). New York: Plenum Press, 1967.

18] Mannan SK, Ruddock KH and Wooding DS Fixation sequences during visual examination of briefly presented 2D images. Spatial Vision 1997, 11, 157-78.

19] Renninger LW, Verghese P and Coughlan J Where to look next? Eye movements reduce local uncertainty. Journal of Vision 2007, 7, 1-17.

20] Najemnik J and Geisler WS 2005 Optimal eye movement search strategies in visual search. Nature, 2005, 434, 387-391.

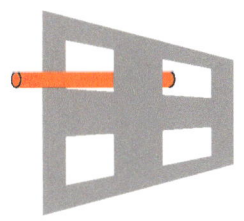

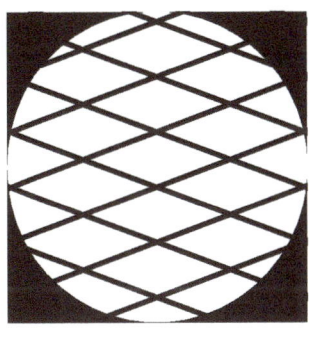

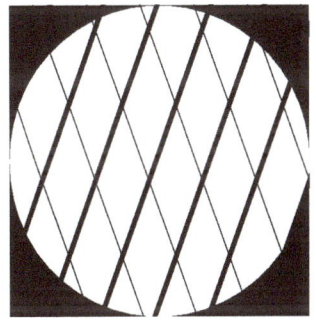

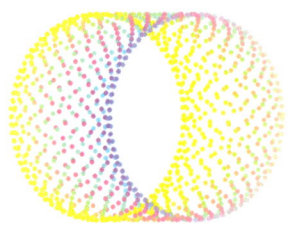

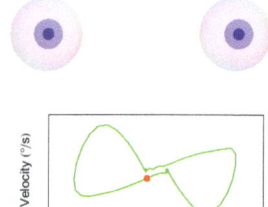

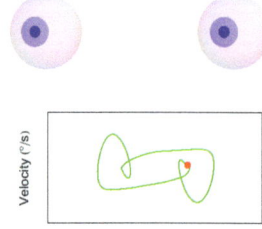

4.0 SEEING SPACE

4.1 Dimensionality reduction

Although the retinal image of an object can change in many different ways the object itself is usually quite limited in the way it can change. For example, a rigid object has only six degrees of freedom; three to specify the translation of the object to its current position and three to specify the rotation of the object to its current orientation. So the position of a rigid object can be specified by a point in a six dimensional space, and movements of the object form trajectories in the six dimensional space. The collection of all possible trajectories is an example of a manifold, which is the n-dimensional generalisation of a surface [1]. In order to be able to interact with an object the visual system somehow has to recover the structure of the manifold which describes how the object can change.

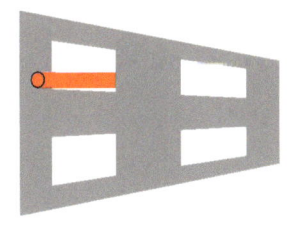

At the retinal level this task is problematic because the six dimensional smooth manifold of an object is embedded in a higher-dimensional space in which each point corresponds to the instantaneous firing rate of a million or so retinal neurons. The situation is worse in the visual cortex which has 10 million neurons or more. The problem of recovering the form of a low dimensional manifold when is embedded in a high-dimensional space is common in data analysis tasks. Every data record consisting of n numbers can be represented by a weighted sum of a set of n base vectors ($\mathbf{u}_1, \mathbf{u}_2, \ldots \mathbf{u}_n$):

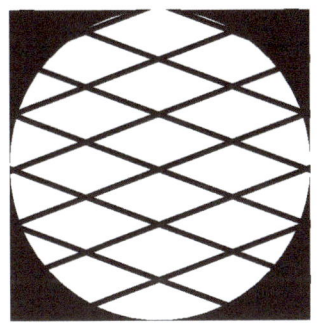

$$\mathbf{x} = x_1\mathbf{u}_1 + x_2\mathbf{u}_2 + \ldots + x_n\mathbf{u}_n$$

A dimensionality reduction transformation is a change of basis to an m-dimensional space which can be used to generate an approximation $\underline{\mathbf{x}}$ of every point $\mathbf{x}$ in the original space.

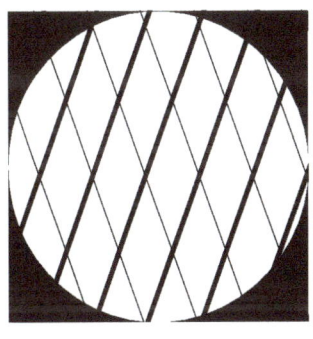

$$\underline{\mathbf{x}} = x'_1\mathbf{u}'_1 + x'_2\mathbf{u}'_2 + \ldots + x'_m\mathbf{u}'_m$$

Note that in this case although the space is m-dimensional the individual basis vectors have n elements, so that the estimated data records and the actual data records have the same length [2].

The simplest form of dimensionality reduction is based on the idea that the values of many of the coordinates in the n-dimensional space are correlated, because they depend on only a few underlying variables. The first step in approximating the original data consists of finding a weighted sum of the original basis vectors such that the average of the differences between the actual data and the approximation are is as small as possible. The next step consists of finding another set of weights which does the same thing for the remaining differences. The process can be repeated until the discrepancies between the actual data and the estimated data is acceptably small.

An example of a dimensionality reduction calculation is the task of representing the reflectance functions of a set of coloured objects. A useful test is provided by the Munsell chip set, which is a set of coloured chips which have been selected so that the changes between neighbouring chips of hue, saturation and lightness appear

equal. A schematic colour diagram of the colours with two different levels of brightness are shown in figure 4.1. Measurements of the reflectance functions have been made at five nanometre intervals in the range 300 to 700 nanaometres, so each digitised reflectance function specifies a pont in a seventy-four dimensional space [3]. The dimesionality reduction technique shows that three weighted sums of these reflectance functions can provide a serviceable approximation for the majority of the colours of the chip set [4].

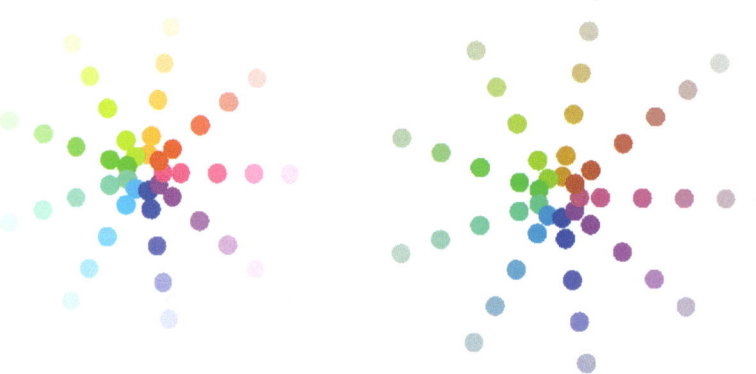

Figure 4.1 Schematic illustrations of the colours of two subsets of the Munsell chip set. Hue varies with polar angle and the saturation with radial position. The two charts correspond to two different lightness levels.

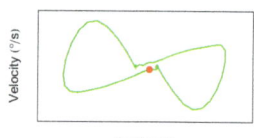

The dimensionality reduction task is well within the capability of the visual system because the required weights can be found by a neural network which follows a Hebbian learning rule. According to this rule the weight associated with an input fibre of a neuron increases if the input fibre and the output fibre are active together. If the change in weight w_i associated with an input x_i is denoted by Δw_i then the Hebbian rule can be described by the equation :

$$\Delta w_i = \mu \, x_i \, y$$

where μ is the learning rate. As it stands this rule can lead to unstable behaviour because the weights can grow infinitely large, but this situation can be avoided by scaling the weights so that the length of the weight vector is always one [5]. Furthermore, the learning rule can be extended to networks so that weights vectors of successive outputs correspond to the successive weight vectors required for the dimensionality reduction transformation [6].

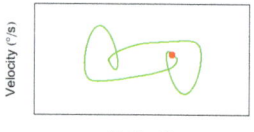

4.2 Colour space

Many light sources can appear to have the same colour, despite having different wavelength distributions. This phenomenon occurs because there are only three types of cone, each of which responds best to a different wavelength. The results of colour matching experiments can be directly related to the properties of the cones, by expressing the results in a space spanned by the spectral sensitivity functions of the cones. This vector space is referred to as the **lms** space, corresponding to the three classes of cones with peaks in their spectral sensitivities at long, medium and short wavelengths respectively. Estimates of the spectral sensitivities of the cones were initially derived from colour matching experiments with subjects who lacked one of the cone types [7,8] and were corroborated by subsequent direct physiological measurement of the cone sensitivities. Recent estimates of the spectral sensitivity curves of the cones are shown in figure 4.2 [9]. The ratios of the maximum sensitivities of the l, m and s signals are 1: 0.5 : 0.025 respectively, and these ratios reflect the relative number of the different types of cone in the retina [10].

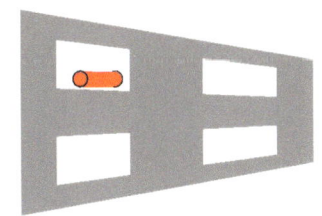

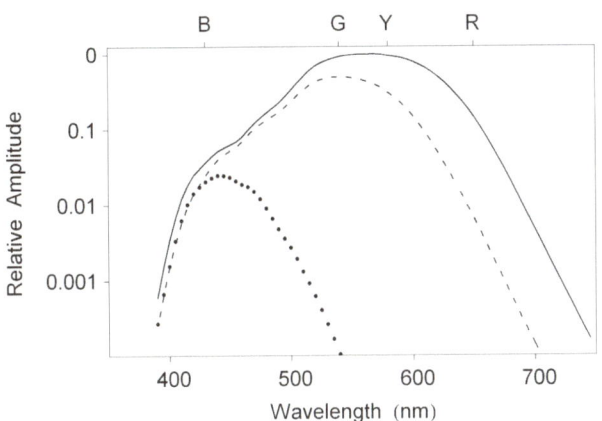

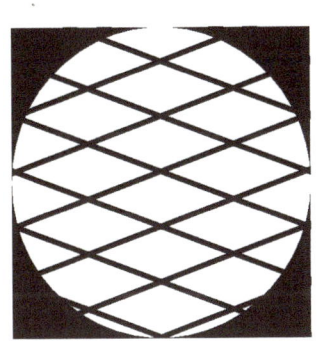

Figure 4.2 Cone spectral sensitivities of the l (continuous line), m (dashed line) and s (dotted line) cones. B, G, Y and R correspond to the wavelengths which appear blue, green, yellow and red respectively.

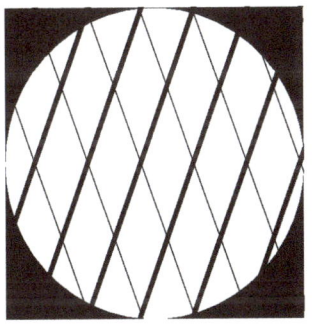

The neural representations of the Munsell chip set can be computed by applying three filters with spectral tuning characteristics corresponding to those of the cones to the reflectance functions of the chips. If the spectral sensitivity function of one channel is s[λ] and the reflectance function of the Munsell chip is m[λ], then the response r of the cone is given by the integral of s[λ]m[λ] over all λ. This function depends on the overlap of the cone sensitivity and reflectance functions, being zero if there is no overlap and maximal if the functions are identical. Dimensionality reduction again reveals three basis vectors which can be used to appoximate the cones responses. The major part of the approximation is provided by a light-dark function of wavelength, the next largest part of the approximation is provided by a red-green function of wavelength and the third part of the approximation is provided by a blue-yellow function of wavelength. These are plotted in figure 4.3 [11].

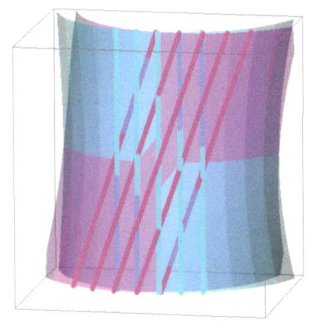

Although the assumption of cone numbers in the ratio l:m:s equal to 1:0.5:0.025 is valid for an average over many subjects, individuals show great variability in their cone ratios, with l:m ratios ranging between 10:1 and 1:2. If the opponent colour channels were fixed linear combinations of the cone signals, then one would expect the wavelength corresponding to the percept of unique yellow, which appears neither reddish or greenish, to vary with the cone ratios [12]. However experimental investigations have revealed no correlation between l:m ratios and the wavelength of the unique yellow, which varies little from 580 nm [13]. The conclusion drawn from these experiments was that a mechanism of neural plasticity is involved in setting the relative weights of the cones in the opponent colour channels. The Hebbian learning mechanism which underlies the dimensionality reduction process explains the invariance of the unique yellow, because the predicted zero crossings of the red-green basis vector with l:m ratios of 10:1, 2:1 and 1:2 are at wavelengths of 576, 578 and 578 nanometres respectively.

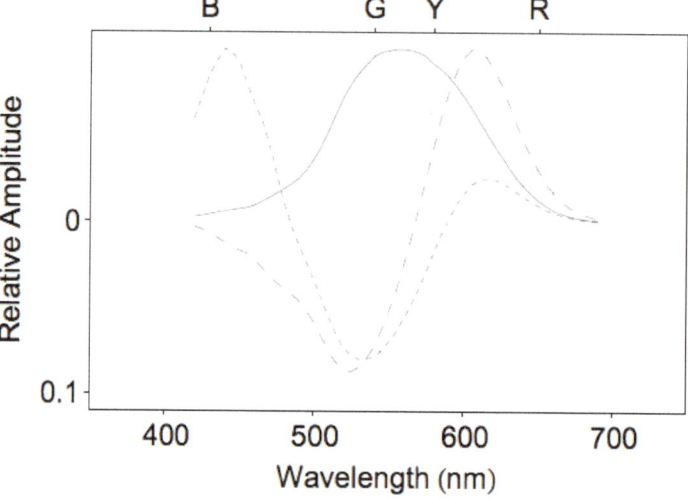

Figure 4.3. Basis vectors for approximating cone responses to the Munsell chip set. The light-dark function is plotted as a continuous line, the red-green function is plotted as a dashed line and the blue-yellow function is plotted as a dotted line.

The dimensionality reduction reveals an opponent colour coding scheme which is in accord with subjective descriptions of colour appearance, as illustrated in figure 4.4. One may see a greenish blue, but never a yellowish blue, and similarly one may see a yellowish green, but never a reddish green [14].

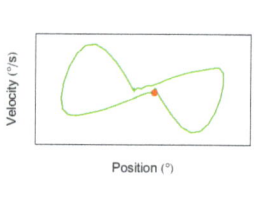

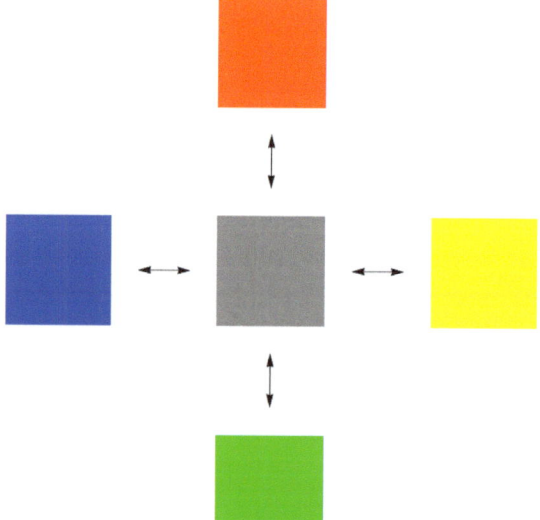

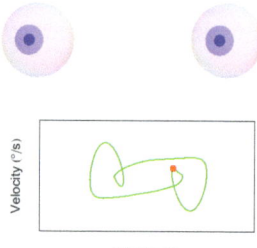

Figure 4.4 Opponent colour sensation axes.

Although the trichromatic theory explains the appearance of uniform patches of light, it falls down as soon as the surface being looked at is not uniform. The simplest example of such simultaneous contrast effects, where the perceived colour of a patch of light depends on the neighbouring patches, is provided by a grey square with black and white backgrounds, as shown in figure 4.5. The same grey square looks darker with the white background and lighter with the dark background [15].

Figure 4.5 Examples of colour contrast effects.

4.3 Spatial frequency space

Both single unit studies of the responses of visual cells to drifting sinusoidal gratings and psychophysical studies on adaptation to gratings have revealed the existence in animal visual systems of filters which are tuned to particular spatial frequencies. In humans, the characteristics of the filters change from having a bandwidth of more than 2 at 0.5 cycles per degrees, to a bandwidth of less than 1 at 20 cycles per degree. The spatial frequency tuning characteristics of the channels can be fitted by exponential functions [16] and the characteristics of one set of channels is plotted in figure 4.6.

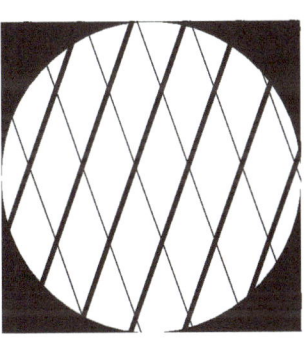

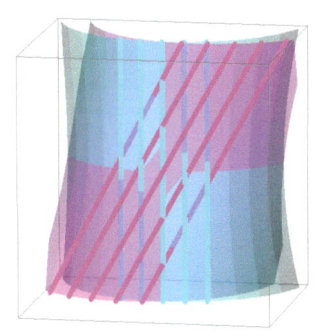

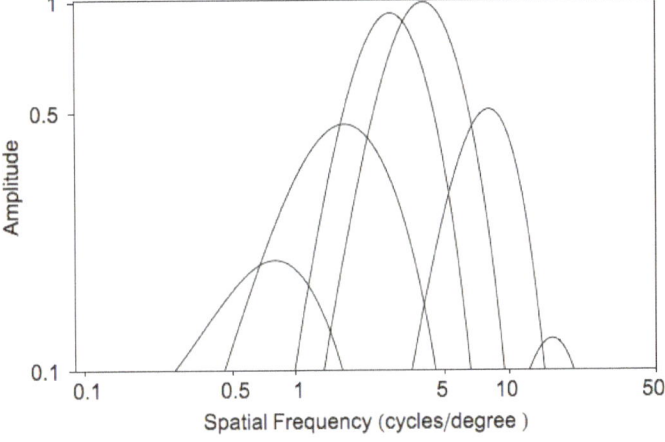

Figure 4.6 Human spatial frequency channels

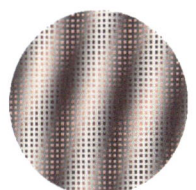

Just as with the dimensionality reduction of the cone signals made in response to the Munsell chip set, dimensionality reduction of the spatial frequency channel signals in response to a database [17] of natural images reveals an opponent spatial frequency organisation. The first two channels are illustrated in figure 4.7.

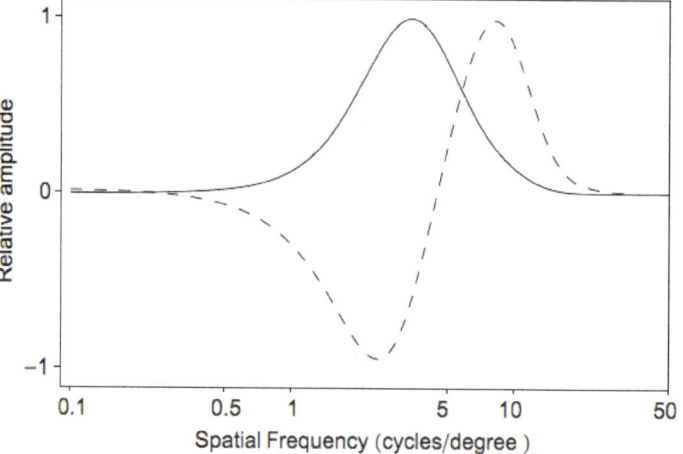

Figure 4.7 Opponent spatial frequency channels

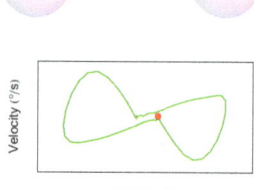

One of the earliest arguments in favour of the opponent colour coding scheme, was that it was in keeping with the subjective appearance of colours. If there an analogous coding scheme for spatial vision, then one would expect to find similar correspondences with the subjective appearances of spatial stimuli. By analogy with the colour vision scheme, the first spatial filter will correspond to a contrast channel, whilst the second spatial filter will correspond to an opponent frequency channel, which signals whether the local texture contains high or low spatial frequencies as illustrated in figure 4.8. Because of the greater variance carried by the contrast channel, visual effects associated with this channel are much more striking than those associated with opponent frequency channel.

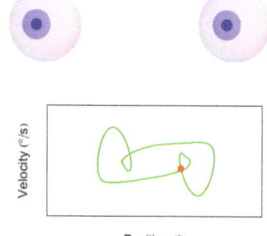

Figure 4.8 Opponent spatial frequency sensation axes.

Simultaneous contrast effects reveal parallels between the perception of apparent contrast and apparent spatial frequency [18]. These parallels are illustrated in figure 4.9. Each square in the figure is designed to subtend a 2 degree by 2 degree patch of the visual field. The upper squares then have the correct spatial frequency to optimally stimulate the contrast channel. The central portions both have the same contrast but simultaneous contrast effect results in the central portion on the right appearing to have a higher contrast. The response of the opponent frequency channel to these stimuli will be close to zero but the response of the contrast channel will show peaks at the borders of the central and outer portions of the squares. The lower squares have central portions are designed to optimally stimulate the opponent frequency filter. In this case the central portion appears to have a higher frequency on the left than on the right. The response of the contrast channel is approximately constant across both squares.

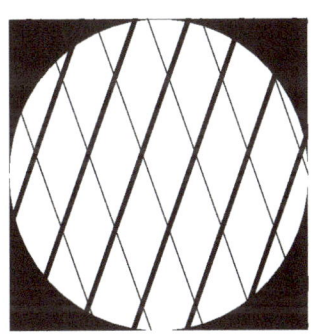

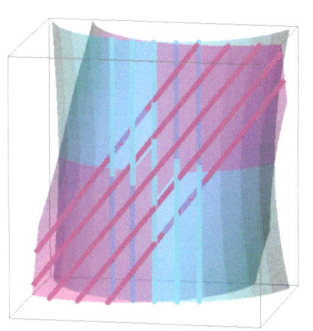

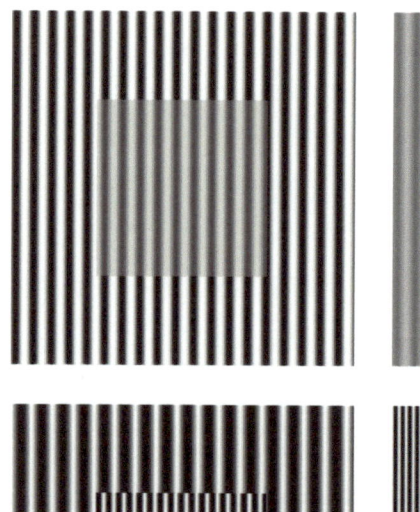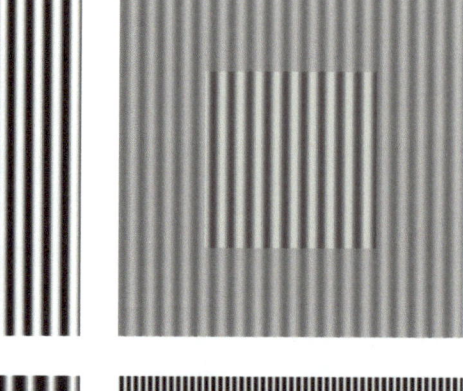

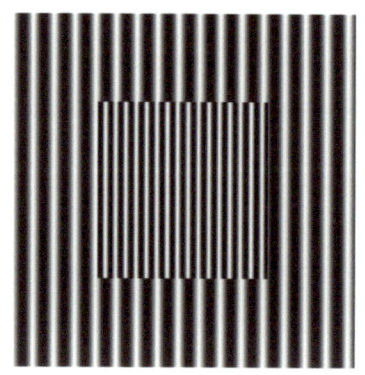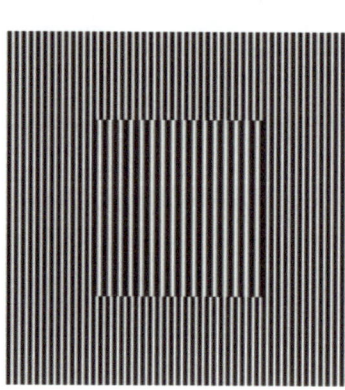

Figure 4.9 Examples of spatial contrast effects

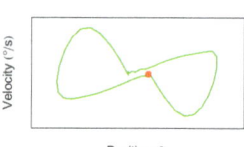

4.4 Eye movement space

The counterpart of the dimensionality reduction of sensory signals is the dimensionality increase of motor signals when going from the central nervous system to the periphery. Amongst the simplest examples of motor signals are those involved in movement of an eye, because only six muscles are involved.

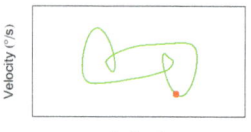

The eye is embedded in fatty and connective tissue so that the movements which it makes are predominantly rotational. In order to specify the orientation of the eye one has to specify not only the direction in which the eye is looking but also the rotation of the eye about that direction. The direction in which the eye is looking can be specified by the line of fixation, which passes through the centre of the fovea and the centre of rotation of the eye. The direction of the line of fixation and the rotation of the eye around it can be specified by a set of three angles, referred to as Euler angles. The first two angles specify the orientation of the line of fixation and the third specifies the rotation of the eye about the line of fixation. The Euler angles are defined by a sequence of rotations which transforms one set of Cartesian basis vectors into another. One of the sets (**x**, **y**, **z**) is fixed in space and the other (**x'**, **y'**, **z'**) is fixed with respect to the rotating object. For eye movements, the set (**x**, **y**, **z**) is fixed with respect to the head and the set (**x'**, **y'**, **z'**) is fixed with respect to the eye. The head based system of axes is orientated so that the **z** direction lies along the straight ahead direction of the line of fixation, the **x** basis vector lies in a horizontal plane and the **y** basis vector is lies in a vertical plane. When the eye is in the straight ahead direction the two sets of base vectors are coincident.

The first rotation in the sequence which defines the Euler angles is a clockwise rotation around the z axis through an angle α as illustrated in figure 4.10A. This angle specifies the meridian in which the line of fixation will end up after the sequence of rotations has been completed and has a range $0 \leq \alpha \leq 2\pi$. The second rotation in the sequence is a clockwise rotation around the new direction of the **x** axis through an angle β as shown in figure 4.10B. This angle specifies the angle that the line of fixation will end up making with the straight ahead direction and has a range $0 \leq \beta \leq \pi$. The final rotation is a clockwise rotation around the new direction of the **z** axis through an angle γ as pictured in figure 4.10C This angle has a range $0 \leq \gamma \leq 2\pi$.

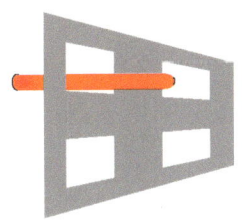

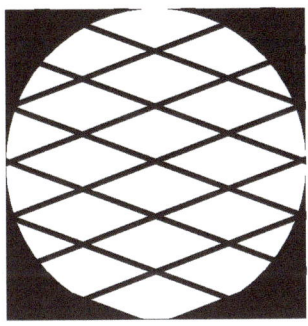

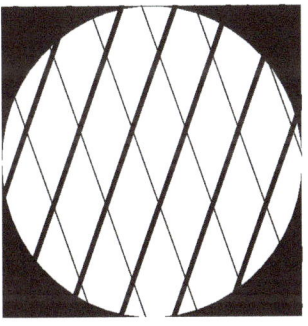

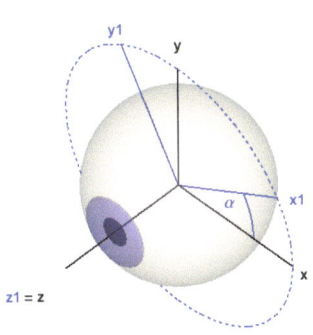

Figure 4.10. Illustration of the relationship between the two sets of Cartesian basis vectors used in the definition of the Euler angles. A. The first rotation in the sequence which defines the Euler angles is a clockwise rotation around the head-fixed **z** axis through an angle α.

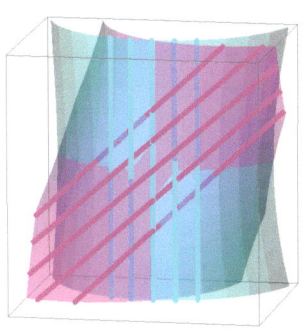

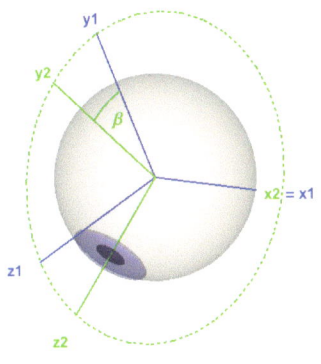

Figure 4.10. Continued. B. The second rotation in the sequence is a clockwise rotation around the new direction of the **x** axis through an angle β.

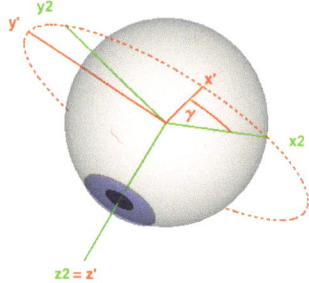

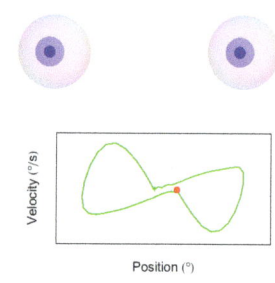

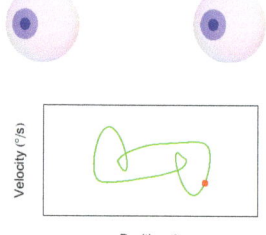

Figure 4.10. Continued. C. The final rotation is a clockwise rotation around the new direction of the **z** axis through an angle γ.

The process of maintaining the image of an object of interest centred on the fovea is referred to as fixation. The rotation can be investigated by using afterimages, which have a fixed location on the retina and so move with the eye. A straightforward procedure involves forming an afterimage of a cross, with the eyes in the straight ahead position looking at a far point on the horizon with the head erect, and then measuring the apparent orientation of the cross whilst looking at a wall perpendicular to the primary position, as shown in figure 4.11. The cross is not distorted when the point of fixation lies on the horizontal or vertical meridian. In general, an afterimage of an oblique line does not appear to alter its orientation at points along the associated oblique meridian, so the axis of rotation must be perpendicular to the oblique meridian. This finding is summarised by Listing's law which states that if the eye moves about a centre O so that the line of fixation moves away from the straight ahead position OA to another position OB, then the displacement of the eyeball is equivalent to rotating it around an axis perpendicular to the plane AOB [19]. When expressed in terms of Euler angles, this law takes the simple form that for every position of the eye α = - β.

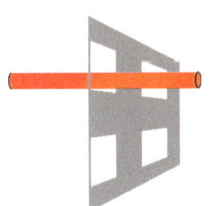
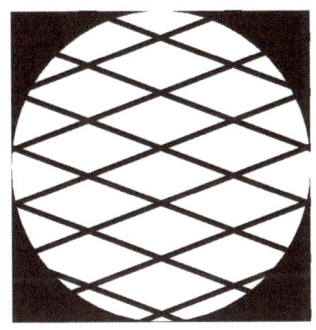

4.11 Figure Procedure for investigating the orientation of an eye using afterimages. The afterimage of a cross is formed with the eye in the straight ahead position. When the eye is subsequently turned to look at various gaze directions, the cross appears distorted in diagonal positions.

An m-dimensional smooth manifold locally looks like the space defined by an ordered set of m numbers $(x_1, x_2 \ldots x_m)$. To obtain a unique set of coordinates for every point on the manifold it must usually be embedded in a higher-dimensional space defined by an ordered set of n numbers $(x_1, x_2 \ldots x_n)$ where n>m. Examples of one dimensional manifolds are an infinite straight line, which can be embedded in the one-dimensional space (x_1) and a circle, which can be embedded in the two-dimensional space (x_1, x_2). Examples of two dimensional manifolds are the unbounded plane which can be embedded in the two-dimensional space (x_1, x_2) and the surface of a sphere, which can be embedded in the three-dimensional space (x_1, x_2, x_3). The embedding is equivalent to having a unique hyperplane at each point on the manifold, with the dimension m of the hyperplane being equal to the dimension of the manifold. The surface of a cone is an example of a two dimensional surface which is not a smooth manifold because there is not a unique hyperplane at its apex.

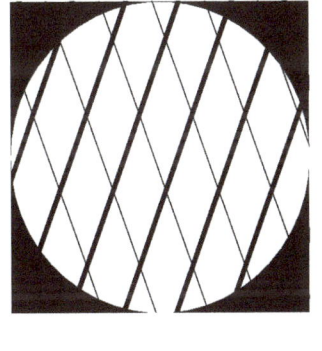
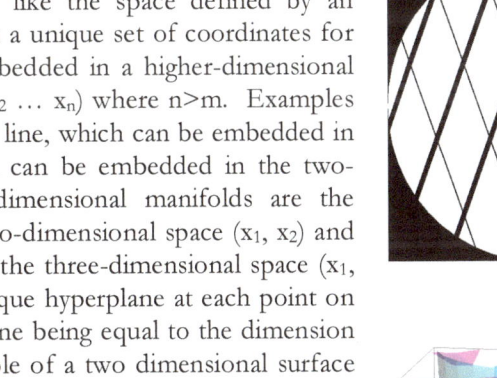
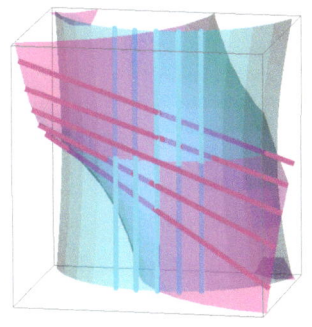

Although the set of possible rotations of the eye defines a three-dimensional manifold, it cannot be embedded in the space defined by an ordered sets of three numbers (x_1, x_2, x_3), so three global coordinates are not enough to specify every possible rotation of the eye uniquely. The motor mechanism has sufficient degrees of freedom to specify every possible rotation uniquely because at least six coordinates are used to specify the innervational levels of the six extraocular muscles, but in the brainstem the motor commands are separated into horizontal, vertical and torsional components. Either there are other components used to increase the dimensionality of the rotational command, or the brain limits itself to representing only a portion of all the possible rotations. Given the physical restrictions on the eye rotations which can be made without damage to the optic nerve the latter alternative seems likely. This point harks back to the beginning of the chapter where the case was made for sensory processing involving recovery of low dimensional manifolds in high dimensional data. Whilst this approach is a useful description of what the nervous system has to do, there is still everything to play for when it comes to defining exactly how such manifolds are represented in the nervous system.

4.5 Visualisation

Sometimes one sees only a part of a shape and has to imagine the entire shape. This is particularly challenging for the shape of space because we see so little of the entire shape. Some of the problems one runs into in imagining the shape of space can be appreciated from the simpler problem of imagining the shape of the surface of the earth. A person standing in the middle of an arid plain can see in every direction. Locally, the world appears to be flat. Nevertheless, the entire surface of the world is that of a sphere. So to imagine the two dimensional surface of the world on has to think of the surface of a three dimensional object. Extending this argument to everyday space implies that one has to think of space as covering some four or more dimensional object.

Unfortunately our ability to see objects appears to be restricted to three dimensional objects. In figure 4.12 the four dimensional extension of a cube is specified, and a stereogram of the projection of the object into three dimensions is shown, and yet it remains resolutely 3 dimensional.

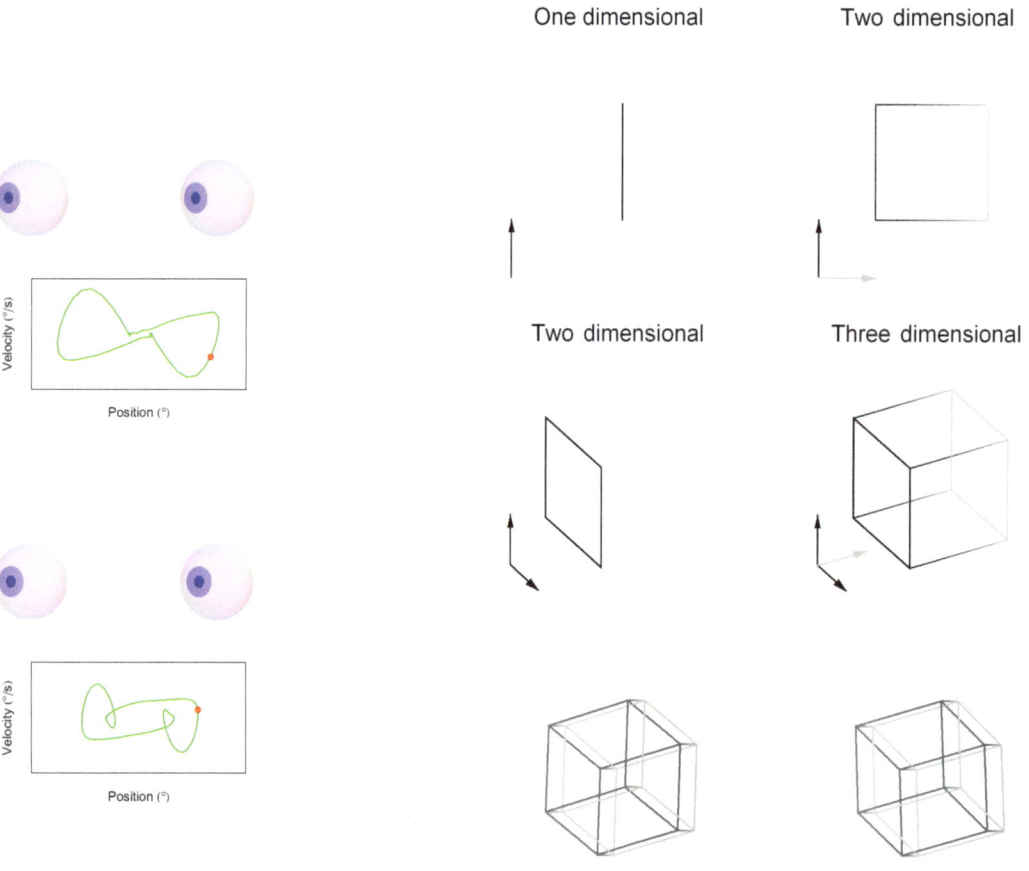

Figure 4.12. Specification of the four dimensional extension of a cube. A vertical straight edge is translated horizontally to sweep out a square. The square is translated in depth to form a cube. Adding another direction enables one to translate the cube in the fourth dimension to specify the four dimensional extension of a cube.

Instead of trying to imagine the surface of a four-dimensional object, an alternative approach is to find ways of imagining the surface independently of the object which it covers. This approach can be emphasized by referring to the surface divorced from its object as a manifold. An example of a two-dimensional manifold known as the torus is provided by the video game in which a representation of a ball on the screen moves in a straight line until it hits the edge of the screen whereupon it reappears in the same position on the other side of the screen. A diagram of the torus can be made by drawing a rectangle and marking opposite sides with arrows which denote how opposite sides are attached to each other. Analogous diagrams of other two dimensional manifolds are shown in figure 4.13.

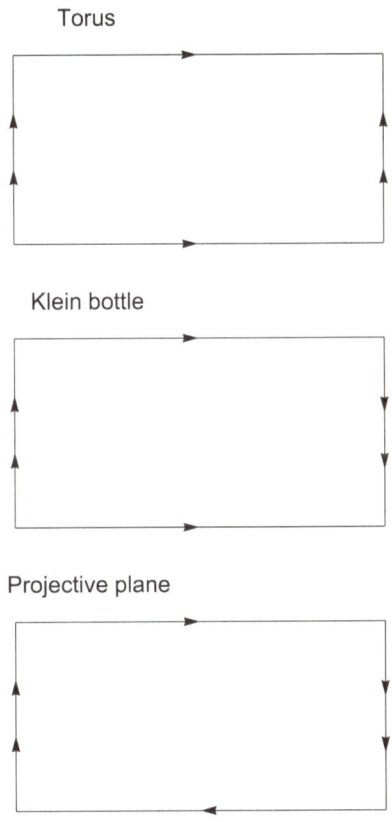

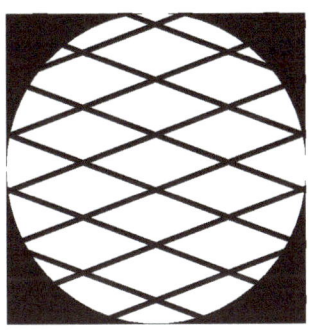

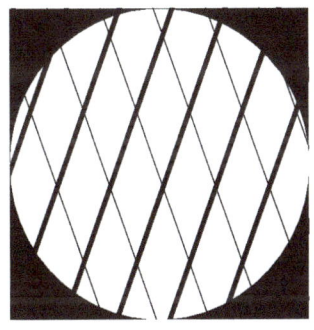

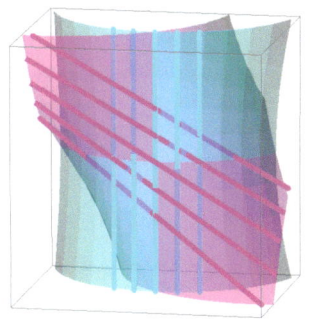

Figure 4.13. Diagrams of some two dimensional manifolds.

The plane diagram of the torus can be drawn on a rubber sheet which can then be physically rolled up and stretched in three-dimensional space to produce a doughnut-shaped object. Hence the torus is topologically equivalent to the surface of a doughnut in the sense that it can be pulled and stretched into the same shape. But it is not geometrically equivalent, because of the distortion caused by the pulling and stretching. In four dimensions one can make a surface which is geometrically equivalent to a flat torus by rolling up the plane diagram horizontally in two dimensions and vertically in the other two dimensions. One way of attempting to visualize this structure is to represent 2 dimensions spatially (let-right and up-down) and 2 dimensions by colour (red-green and blue-yellow). This is illustrated in flip book six, where a circular moving pattern on the plane diagram is shown in four dimensions.

The projective plane is equivalent to the set of orientations of the eye made in accordance with Listing's law. For every rotation away from the straight ahead direction one can define a pair of numbers which specify a point on a plane such that the direction of the point with respect to the origin is equal to the axis of rotation and the distance of the point from the origin is equal to the clockwise angle of rotation in the range 0 to 180 degrees. For rotations through 180 degrees this process gives two diametrically opposite points. If these diametrically opposite points are glued together then a surface equivalent to the projective plane is obtained. Flip book seven is an animation a circular moving pattern on the projective plane diagram embedded in four dimensions.

4.6 References

1] Zhang K and Sejnowski TJ A theory of geometric constraints on neural activity for natural three-dimensional movement. Journal of Neuroscience 1999, 19, 3122-3145.

2] Kirby M Geometric Data Analysis. John Wiley and Sons, Incorporated, New York, 2001.

3] Parkkinen JPS, Hallikainen J and Jaaskelainen T Characteristic spectra of Munsell colors. Journal of the Optical Society of America 1989, 6, No. 2, .318-322.

[http://www.cs.joensuu.fi/~spectral/databases/download/munsell_aotf.htm]

4] Cohen J Dependency of the spectral reflectance curves of the Munsell color chips. Psychonomic Science 1964, 1, 369 –370.

5] Oja E A simplified neuron model as a principal component analyser. Journal of Mathematical Biology 1982, 15, 267-273.

6] Sanger TD Optimal unsupervised learning in a single-layer linear feedforward network Neural Networks 1989, 2, 459-473.

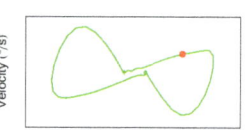

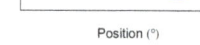

7] Vos JJ and Walraven PL On the derivation of the foveal receptor primaries. Vision Research 1970, 11, 799-818.

8] Smith VC and Pokorny J Spectral sensitivity of the foveal cone photopigments between 400 and 500 nm. Vision Research 1975, 15, 161-171.

9] Stockman A and Sharpe LT The spectral sensitivities of the middle- and long-wavelength-sensitive cones derived from measurements in observers of known genotype. Vision Research 2000, 40, 1711-1737.

10] Bowmaker JK, Dartnall HJA, Lythgoe JN and Mollon JD The visual pigments of rods and cones in the rhesus monkey, *Macaca Mulatta*. Journal of Physiology (London), 1978, 274, 329-348.

11] Buchsbaum G and Gottschalk A Trichromacy, opponent colour coding and optimum colour information transmission in the retina. Proceedings of the Royal Society of London, Series B 1983, 220, 86, 113.

12] Cicerone, C.M. (1987) Constraints placed on color vision models by the relative numbers of different cone classes in human fovea centralis. Farbe, 34, 59-66.

13] Neitz, J., Carroll, J., Yamauchi, Y., Neitz, M. & Williams, D.R. (2002) Color perception is mediated by a plastic neural mechanism that is adjustable in adults. Neuron, 35, 783-792.

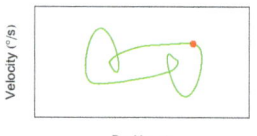

14] Hering E Outlines of a Theory of the Light Sense. Harvard University Press, Cambridge 1964. English Translation by L. Hurvich and D.Jameson (originally published in 1874).

15] Albers J Interaction of Color. Yale University Press, New Haven 1963.

16] Wilson HR Psychophysical models of spatial vision and hyperacuity pp 64-86 in Spatial Vision, Volume 10, Visual Dysfunction, edited by D.Regan. Macmillan Press, London 1991.

17] Van Hateren JH and van der Schaaf A Independent component filters of natural images compared with simple cells in primary visual cortex. Proceedings of the Royal Society of London Series B 1998, 265, 359-366.

[http://hlab.phys.rug.nl/archive.html].

18] Mackay DM Lateral interaction between neural channels sensitive to texture density. Nature 1973, 245, 159-161.

19] Helmholtz H von Treatise on Physiological Optics. 1910. Translated by J.P.C.Southall, New York, Dover, 1924.

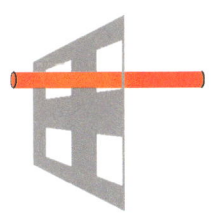

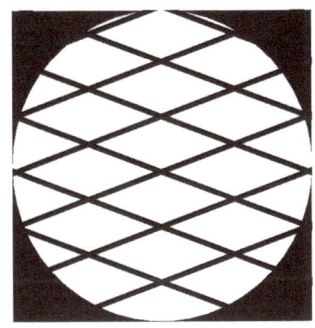

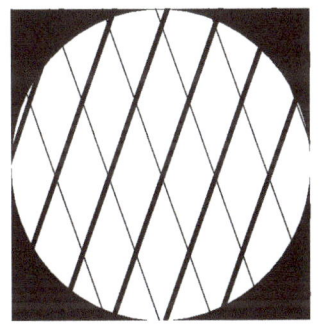

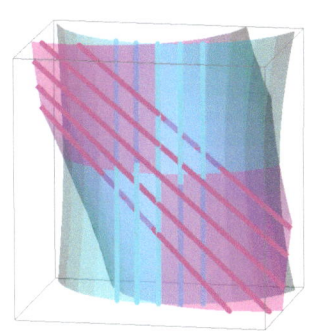

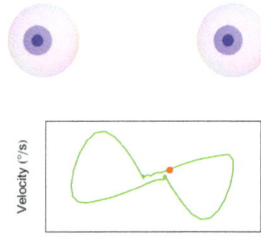

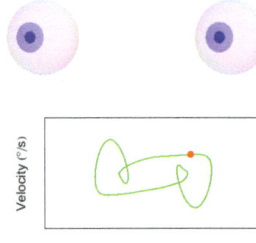

56

5.0 AGAINST BLOCK DIAGRAMS

5.1 Structure and algebra

The computation performed by any neural network involves a transformation of its input into its output. This transformation can be conceptualised by a block with an input coming into it from one direction and an output leaving it from another. Given a block representing a transformation T then the original input can be recovered if is possible to find an inverse transformation T^{-1} which reverses the original transformation. To distinguish it from the inverse transformation the original transformation is referred to as the forward transformation.

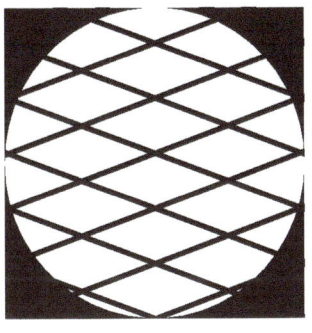

So if one had to design a visual system, one might begin by specifying the forward transformation S that occurs in the optics and sensory transduction and apply the inverse transformation S^{-1} to recover the original picture. This approach is illustrated on the left in figure 5.1. An alternative scheme uses a model of the forward transformation to generate successive guesses at the output from the forward transformation until the difference is minimal. Although this scheme takes longer to recover the retinal image it can be useful because it is sometimes simpler to compute a model of the forward transformation than its inverse.

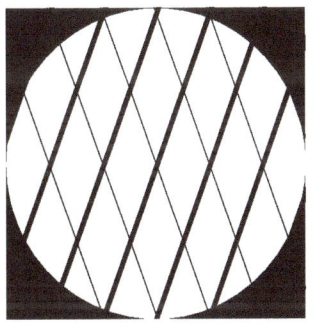

Just as the sensory pathway requires an inverse transformation to recover the original retinal image, so the motor pathway requires an inverse transformation to transform a required movement into an actual movement. For example, the most direct motor pathway for saccades would involve an inverse transformation preceding the motor transformation, as shown in the left half of figure 5.2. But the same result could also be achieved by using a forward transformation within a negative feedback loop, as illustrated in the right half of figure 5.2. The latter scheme is the one actually used by the oculomotor system because subjects continue to make accurate saccades, albeit slower, after they have been given tranquillizing drugs. Without the feedback loop, the saccades would no longer finish on the target. In the saccadic feedback loop, an integrator suffices as an adequate model of the forward transformation and an amplifier is all that is required for the controller [1].

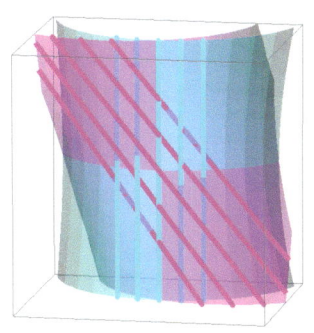

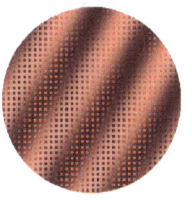

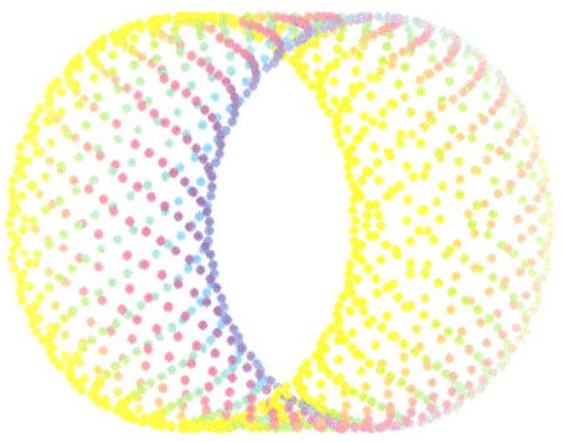

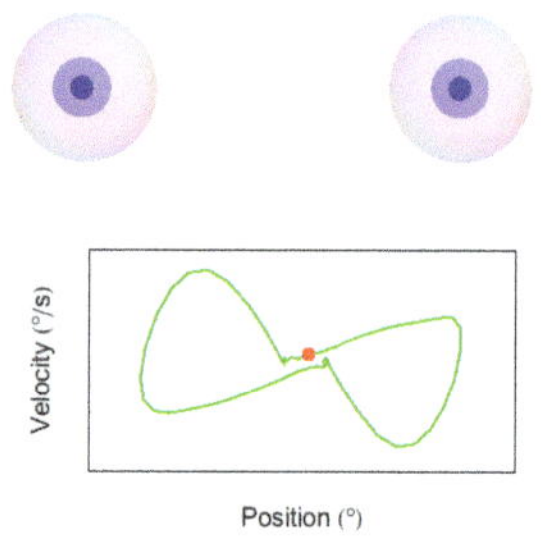

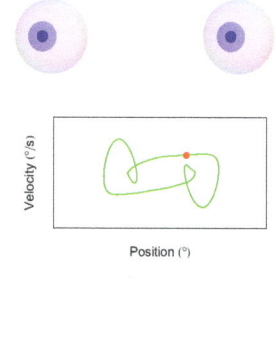

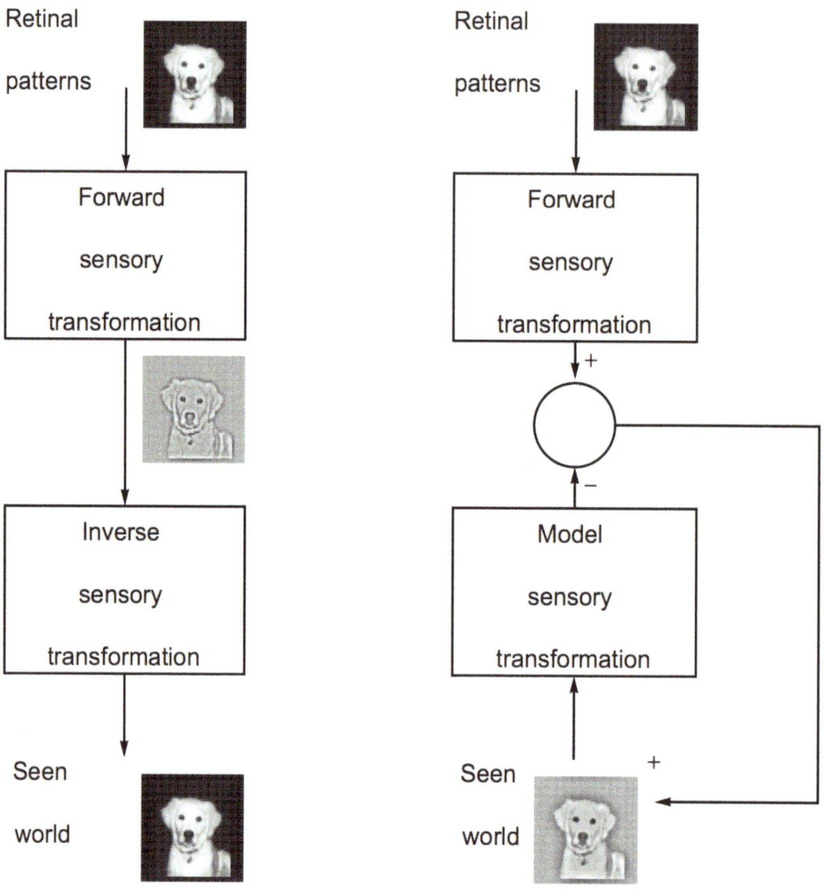

Figure 5.1 Two alternative schemes for processing retinal images.

58

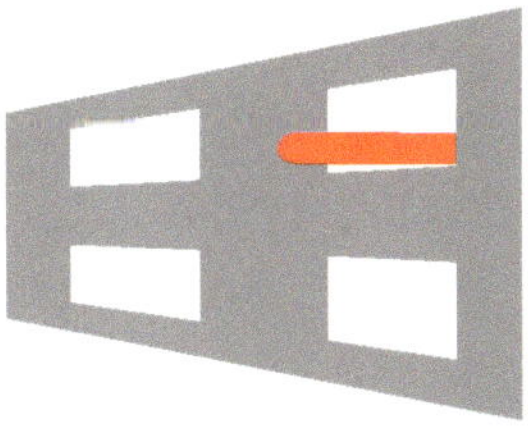

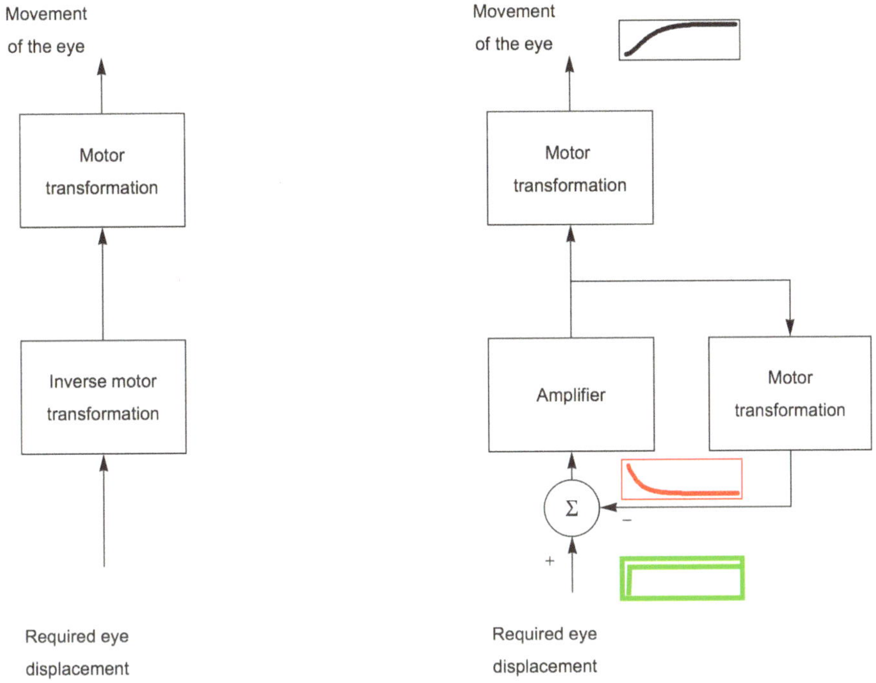

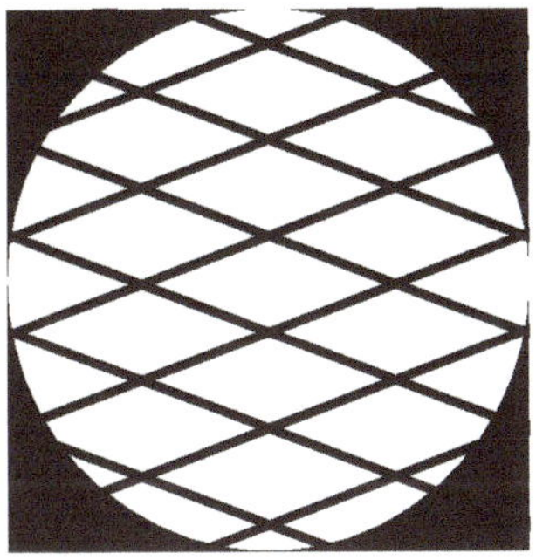

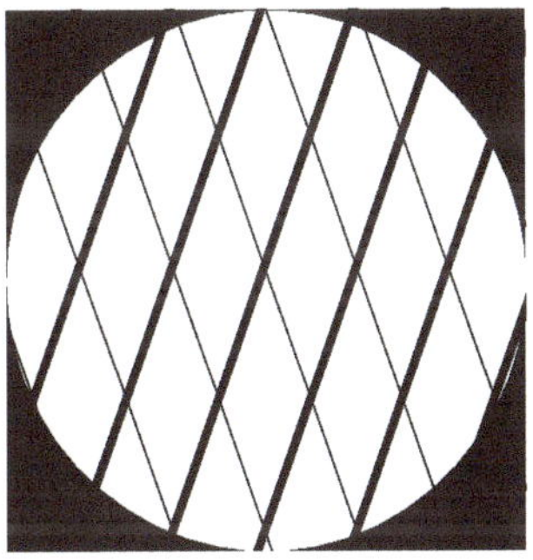

Figure 5.2 Alternative schemes for generating an eye movement command.

The main advantage of considering the transformations of the sensory and motor images is that one can go on to identify the structures in the brain which are responsible for carrying out the transformations. For example, for the usual range of head movements the signal provided by the semicircular canals describes head velocity and the signal provided by the firing of the oculomotor neurons has to ensure that the required tension developed in the muscles overcomes both the spring force which depends on the position of the eye and the damping force which depends on the velocity at which the eye moves. To obtain the signal required to hold the eyes steady after an eye movement made to compensate for a head movement, there must exist a neural mechanism which can convert the eye velocity into eye position [2]. This argument lead to the successful identification of the circuitry in the brainstem responsible for this transformation.

5.2 Behaviour and geometry

The circuitry of the brain contains an enormous number of feedback loops and any network with feedback will continue to reverberate after it is presented with an input.

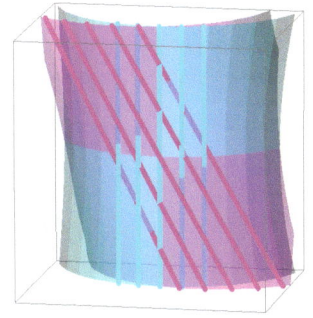
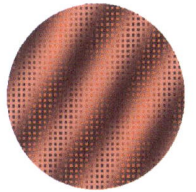

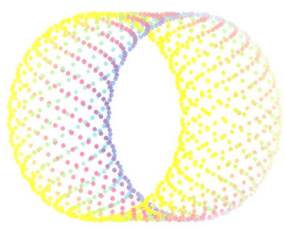

If the input does not change eventually the network will settle down into a stable pattern of behaviour, which typically consists of staying in one state or a repeating cycle of states. A useful tool for understanding the types of processing that a network is capable of is provided by a geometrical description according to which the state of the network is described by a point in a state space and successive states trace out a curve in state space, referred to as a trajectory. An equilibrium state does not change over time and its respresentation in state space is referred to as a fixed point.

The behaviour of the network is determined by the vectors associated with every point in the state space which specify how the network will change from one state to the next. Usually these vectors are organised in the neighbourhood of the fixed points so that trajectories approach the point along paths lying on a stable manifold and move away from the point along paths on an unstable manifold. Networks which exhibit an alternation of slowly varying and rapidly varying behaviours can be modelled by a choosing vectors which move the state of the system rapidly onto a stable manifold whereupon the state moves relatively slowly towards a fixed point[3]

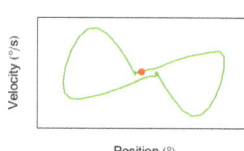

For example, the physiological mechanism underlying the generation of saccades can be considered as a slow-fast system. Saccades show a relatively invariant relationship between the size of the movement and the peak velocity and duration of the movement. The peak velocity of saccades typically varies from thirty to seven hundred degrees per second and their duration varies from thirty to one hundred milliseconds for eye movements of a half to forty degrees in amplitude. The peak velocity progressively saturates with increasing saccade amplitude after 20 degrees. This consistent relationship between amplitude, duration and peak velocity of saccades is termed the main sequence [4].

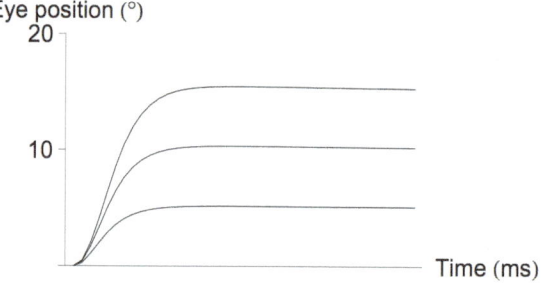

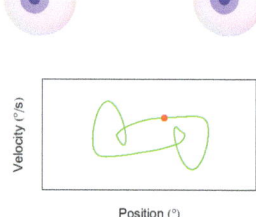

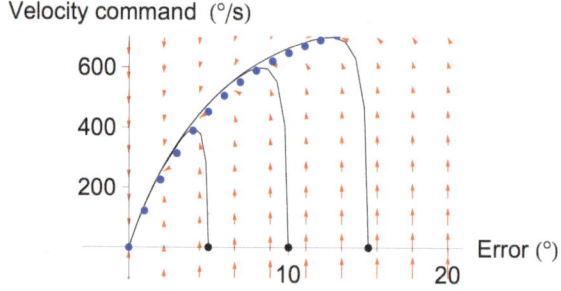

Figure 5.3 Slow/Fast description of the saccade generation mechanism. The state space has two dimensions – the motor error between current eye position and the target eye position and the eye velocity command produced by the burst neurons.

Saccades are generated by a burst of activity in the brainstem which specifies the velocity of the movement. The neural integrator converts this velocity signal to a position signal. These two signals are summed by the oculomotor neurons which drive the extraocular muscles [5,6]. The state of the burst generator can be described by a pair of coordinates; one coodinate specifies the motor error, which is the difference between the current eye position and the target eye position, and the other coordinate specifies the eye velocity command. If the trajectories for different-sized saccades are plotted on this two-dimensional state space then they superimpose on a single curve plotted in blue in figure 5.3 [7]. This is the slow manifold of the system. The arrows on the figure illustrate the vectors associated with each point in the state space. In accordance with the slow/fast structure of the system the vectors act to move the state of the system rapidly onto the slow manifold, and then more gradually along the slow manifold to the stable point attractor at the origin.

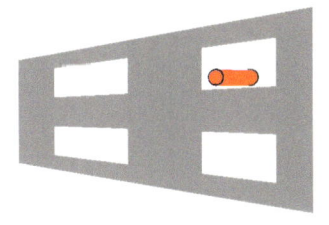

An immediate consequence of the mechanism is that voluntary saccades will follow the main sequence as their trajectories lie along the slow manifold. But the slow/fast mechanism also suggests how the oculomotor disorder of early-onset nystagmus could develop in a mechanism which usually behaves reliably throughout a lifetime[8]. Nystagmus is an eye movement instability consisting of rhythmic too and fro movement of the eyes[9]. When there are separate velocity command generators for movements to the left and movements to the right they will have to be mutually inhibitory when no movement is required to ensure that the overall velocity command is zero. In this case the slow manifold can become distorted at the origin as shown in figure 5.4 and the state of the system can be forced endlessly round a loop in the state space.

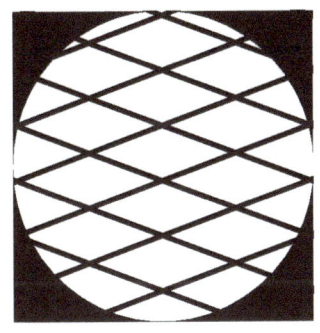

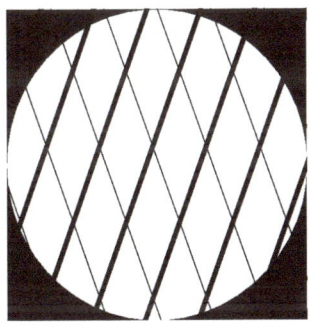

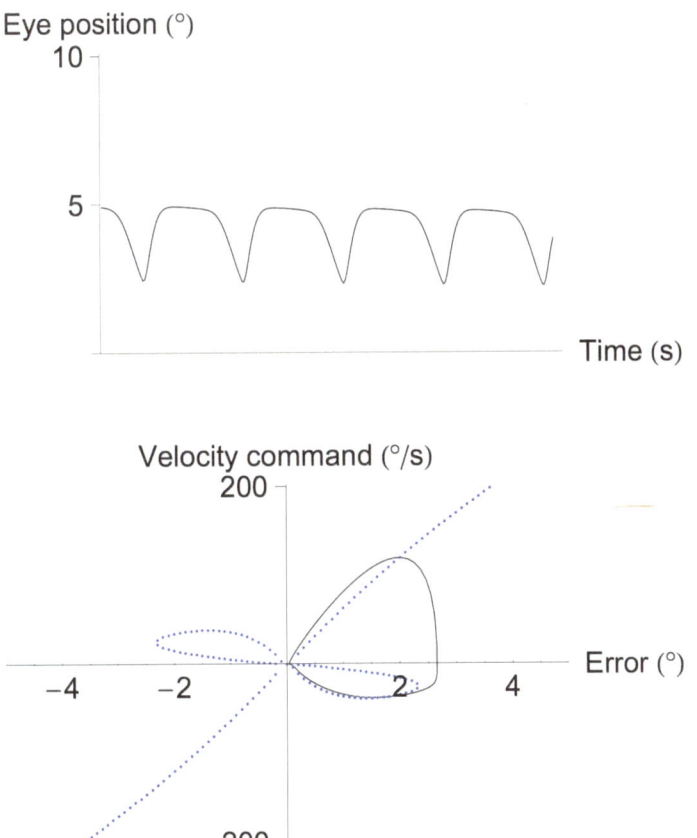

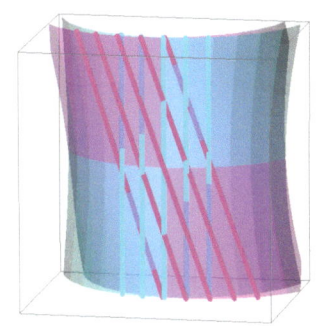

Figure 5.4 Development of an oculomotor disorder by a change in the form of the slow manifold.

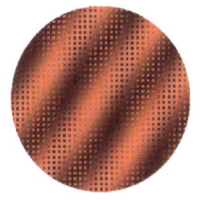

This mechanism explains why sometimes when a person with nystagmus wants to shift their gaze in the direction of the slow phase, then they make the movement not by a saccade but by an extended slow phase as illustrated in figure 5.5. Initiating a saccade requires that the state of the burst generator is displaced along the error direction. If a displacement in the direction of the slow phase is made when the state is at the end of the slow phase then the system will restart the slow phase and a saccade will not be made. However, a displacement in the opposite direction of the slow phase will lead to a larger than normal saccade.

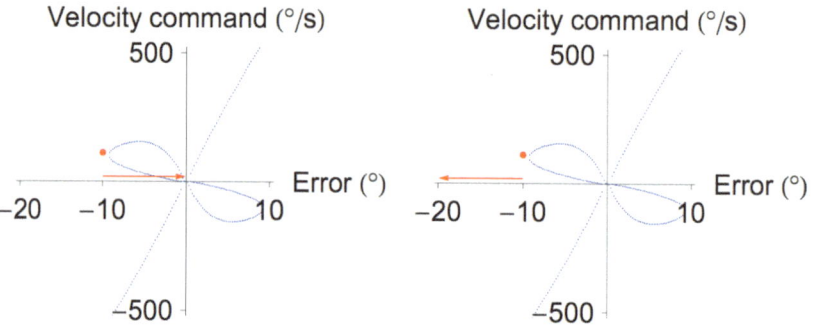

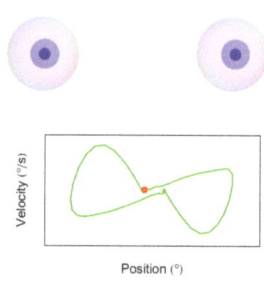

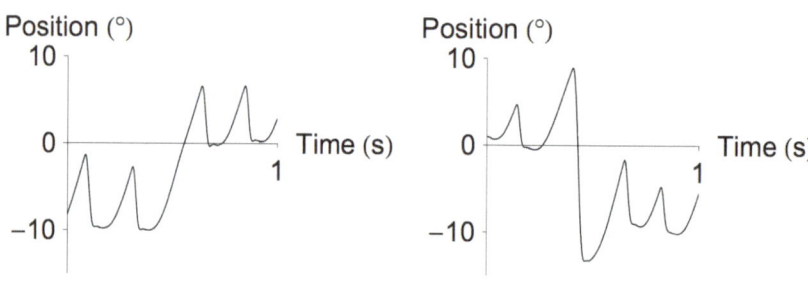

Figure 5.5 Generation of an extended slow phase in nystagmus.

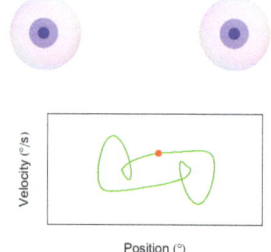

The slow/fast model explains some aspects of nystagmus but it is based on the assumption that there is a single fixed point at the origin whereas experimental data implies that the nystagmus is organized about a pair of fixed points symmetrically placed about the origin. This feature of nystagmus is illustrated in flip book 8, which is an example of bidirectional jerk nystagmus and flip book 9 which is an example of bidirectional pendular nystagmus.

5.3 Albinism and the visual system

The example of nystagmus is useful for highlighting the importance of understanding the range of behaviours which a neural network is capable of. It is hard to explain the ordinary behaviour of the visual system, and so the range of extraordinary behaviours of the visual system are often overlooked. But these behaviours provide important tests of the limitations of our current understanding of the visual system. For instance the condition of albinism leads to many changes in the visual system and we do not yet have a framework in which the variety of these changes can be explained.

Albinism is caused by a reduced level of melanin. In humans, the condition can result from disorders of a number of genes, and for each of these genes, a variety of mutations can occur. The visual system of albinos typically shows a number of distinctive features which include transillumination of the iris, an underdeveloped fovea, abnormal routing of the fibres in optic nerve and nystagmus. The misrouting in albinism is such that the axons of cells in the temporal retina, which would not be expected to cross over at the chiasm, anomalously project to the contralateral side of the brain. This is illustrated in figure 5.6. In effect, the vertical line which marks the decussation boundary is shifted towards the temporal retina. The amount of this shift may be related to the reduction in the level of melanin [10].

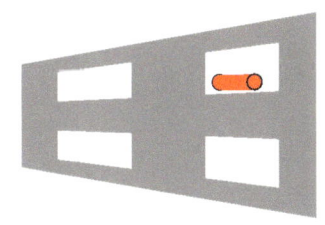

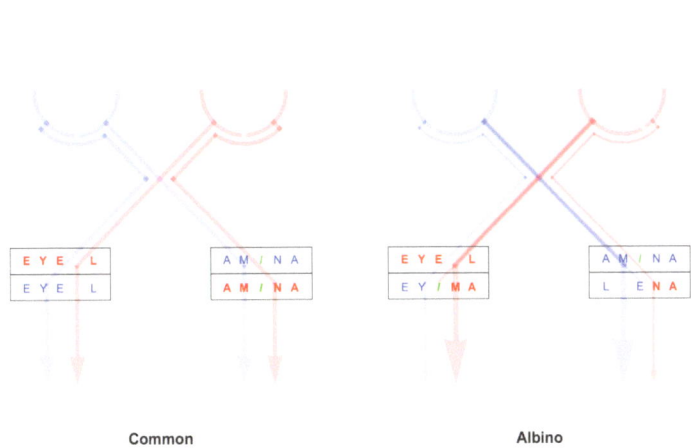

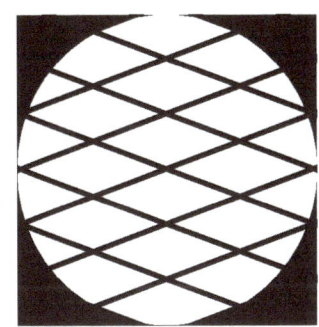

Figure 5.6 Schematic diagram showing the differences between the early visual pathways in subjects with and without albinism. The fibres from the right eye are shown in red and from the left eye in blue. In albinism a larger proportion of fibres project to the contralateral hemisphere. The functional effect of the misrouting is illustrated by the target letters "ANIMAL EYE". After the optical reversal in the eyes, the sequence of these targets becomes "EYE LAMINA". In the normal visual system the projections of these targets onto the monocular layers of the lateral geniculate nucleus are in register. The locations of the representations of the target letters in the monocular layers of the LGN are printed in red for the connections from the right eye and in blue for the connections from the left eye. In the visual pathway of albinos, the ipsilateral projections are altered by the misrouting resulting in the anomalous representation of the sequence of targets "EYIMAL ENA". An implication of this misrouting is that a single target for a saccade, such as the letter I, will have representations in two separate cortical locations which correspond to midline symmetric positions in space in the normal projection.

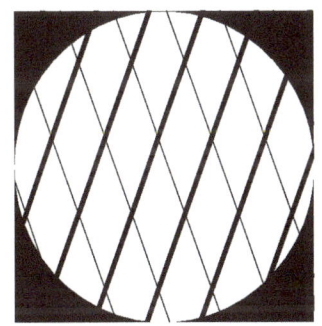

The most prominent difference between the normal and albino visual system is that the sensitivity to visual stimuli is much more variable in the albino visual system. For example the threshold intensity at which the normal observer sees a 1/10 of a degree diameter target on the vertical meridian is lowest at the fovea and increases steeply with retinal eccentricity. Although this pattern can be found in people with albinism, a continuum of other relationships is more commonly found, ranging from less steeply depending on eccentricity through to being independent of eccentricity [11]. On the motor side, although most people with albinism have nystagmus some do not and in these individuals drifts and instabilities of fixation are discernable [12,13]. Our progress in understanding the effects of albinism on the visual system is slow because we do not have a framework for understanding why the manifestations of albinism are so different in different people.

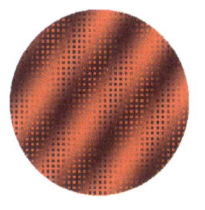

5.4 How does the brain see?

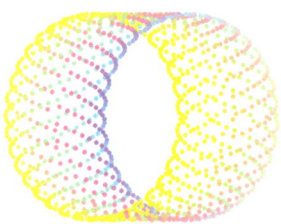

Although all the evidence points to the brain and eye being responsible for seeing, the mechanism involved is still something of a mystery. As with the early stages of understanding any phenomena, the difficulties are primarily conceptual. The review of the theories in chapter 1 highlighted some important properties of seeing which together. Looking back, I seem to have tackled these questions in reverse order.

1] How does the brain guess what would have to be 'out there' to produce the retinal image?

Rather than considering the process of seeing to involve an increase in dimensionality from a two-dimensional retinal image to a three-dimensional world of objects, the process can be taken to require a decrease in dimensionality from the high-dimensional firing pattern of a million or so retinal neurons to the low-dimensional perceptual world of solid objects. We tend to see what is 'out there' because the dimensionality-reduction process is designed to isolate the sources of the changes in the sensory data. Objects are defined by the changes they make in the retinal image and so naturally enough they account for the changes.

2] How does the visual system isolate individual objects in the scene?

Networks make up of spatial arrays of neurons with lateral connections can display a variety of behaviours. The response to an excitatory input can grow until all the neurons are active or diminish until none of the neurons are active. But with the right balance of strengths of excitatory input connections and inhibitory lateral connections activity in the network can persist over many seconds. In the context of this approach to neural networks, memory for objects emerges straightforwardly as a persistent neural activity that is spatially localized [14-17].

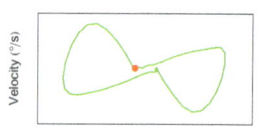

3] How is vision used in movement?

Biological neurons have limited connectivity and limited response ranges so the brain can only carry out mathematical operations with limited precision. The visual and motor systems get round this biological limitation by using population vectors coding to represent movement variables. This representation can implementt sensorimotor transformations by modulating the firing rates in arrays of interneurons which allows position information from hearing and touch to be pooled easily with the visual information.

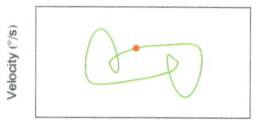

4] What is all the neural activity in different cortical maps representing?

The data which the visual system has to deal with is especially challenging because changes in the retinal image are typically caused by more than one object, which often pass in front of, or behind, one another. Although decorrelation was introduced in chapter 4 as an illustration of a technique for dimensionality-reduction, it will not normally separate out objects which overlap one another. Yet the brain can do this very easily, as can be appreciated from the flip book [5] of the morphing gratings [18]. It is clear that the positions of the parts of the objects are represented topographically and that the maps are somehow superimposed to represent the different properties of the parts, but the details of how this is done are not known. One set of stimuli which are proving very useful for investigating how different properties are represented is provided by simulations of biological movement, because these involve represented of motion in the early visual cortex, solution to the aperture problem in the higher-level visual cortex and subsequent interpretation of the gestures and interactions.[19,20]. An example of this stimulus is shown in flip book 10.

Neurophysiological evidence makes it unarguable that the brains of different animals will represent different properties of objects. The inescapable conclusion is that they are seeing different things 'out there' from the objects seen by humans But even among humans, the objects seen differ between individuals. One wonderful description of the different objects is provided by James Thurber in his short story "The Admiral at the Wheel", which describes the world he sees after his glasses are broken. The immediate goal of modelling the neurophysiological mechanism of vision may not solve any of the world's problems but it promises to reveal the dimensions along which individuals encode the behaviour of objects. Better understanding of how the visual system represents n-dimensional data will enable the design of data visualization tools which are optimally matched to the capabilities of the visual system [21]. Then we may really see the world.

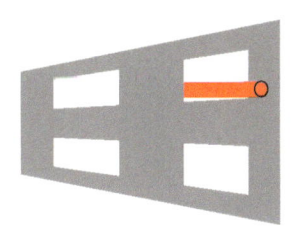

5.5 References

1] Optican LM and Quaia C From sensory space to motor commands: lessons from saccades. Engineering in Medicine and Biology Society. Proceedings of the 23rd Annual International Conference of the IEEE , 2001, 1, 820 -823.

2] Robinson DA Eye movement control in primates. Science 1968, 161, 1219-1224.

3] Zeeman EC Differential equations for the heartbeat and nerve impulse. Pp 8-67 in Towards a Theoretical Biology , volume 4, Edited by CH Waddington. Edinburgh University Press, Edinburgh.

4] Bahill AT, Clark MR and Stark L The main sequence, a tool for studying eye movements. Mathematical Biosciences 1975, 24, 191-204.

5] Scudder CA, Kaneko CRS, Fuchs AF (2002) The brainstem burst generator for saccadic eye movements. A modern synthesis. Exp Brain Res 142: 439-462.

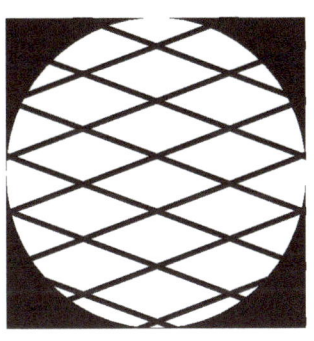

6] Sparks DL (2002) The brainstem control of saccadic eye movements. Nat Neurosci Rev 3: 952-964

7] Van Gisbergen JAM, Robinson DA, Gielen S (1981) A quantitative analysis of generation of saccadic eye movements by burst neurons. J Neurophysiol 45: 417-442.

8] Broomhead DS, Clement RA, Muldoon MR, Whittle JP, Scallan C, Abadi RV (2000) Modelling of congenital nystagmus waveforms produced by saccadic system abnormalities. Biol Cybern 82: 391-399

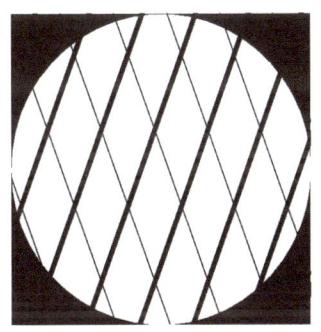

[9] Dell'Osso LF and Daroff RB Congenital nystagmus waveforms and foveation strategy. Documenta Ophtalmolologica 1975, 39, 155-182.

10] Hoffman MB, Tolhurst DJ, Moore AT and Morland AB Organisation of the visual cortex in abinism. The Journal of Neuroscience 2003, 23, 8921-8930.

11] Abadi RV and Pascal E Incremental light detection thresholds across the central visual field of human albinos. Investigative Ophthalmology & Visual Science 1993, 34, 1683-1690.

12] Collewijn H, Apkarian P and Spekreijse H The oculomotor behaviour of human albinos. Brain 1985, 108, 1-28.

13] Timms C, Thompson D, Russell-Eggitt I and Clement R Saccadic instabilities in albinism without nystagmus. Experimental Brain Research 2006, 175, 45-49.

14] Droulez J and Berthoz A A neural network model of sensoritopic maps with predictive short-term memory properties. Proceedings of the National Academy of Sciences of the USA. 1991, 88, 9653-9657.

15] Constantinidis C and Wang X-J A neural circuit basis for spatial working memory. The Neuroscientist, 2004, 10, 553-565.

16] Vogels TP, Rajan K and Abbott LF (2005) Neural network dynamics. Annual Review of Neuroscience 28: 357-76.

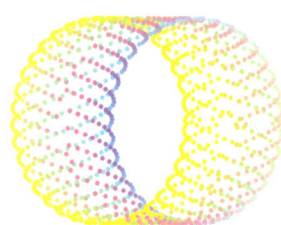

[17] Macoveanu J, Klingberg T and Tegnér J A biophysical model of multiple-item working memory: A computational and neuroimaging study. Neuroscience 2006, 141, 1611-1618.

[18] Blaser E, Plyshyn ZW and Holcombe AO Tracking an object through feature-space. Nature 2000, 408, 196-199.

[19] Cutting JE A program to generate synthetic walkers as dynamic point-light displays. Behavior Research Methods and Instrumentation 1978, 10, 91-94.

[20] Giese MA and Poggio T Neural mechanisms for the recognition of biological movements. Nature Neuroscience Reviews 2003, 4, 179-191.

[21] Ware C Information Visualisation. Second Edition. Elsevier Incorporated, San Francisco, 2004.

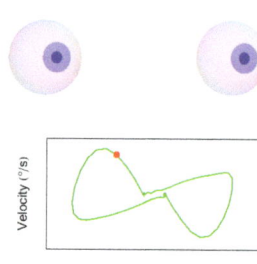

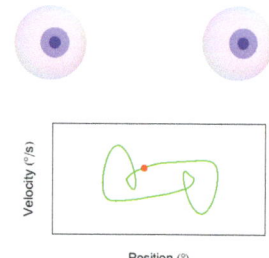

INDEX

"Admiral at the Wheel, The" (Thurber) 65

afterimages 50–1

albinism 62–3

algebra 17–21

Ames window 22

analogue system 11

aperture problem 23

attention 36

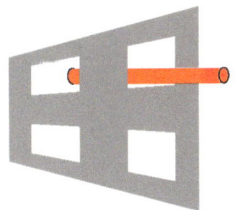

basis vectors 19

Bayes's formula 10–11

biological movement simulation 64

burst generator 61

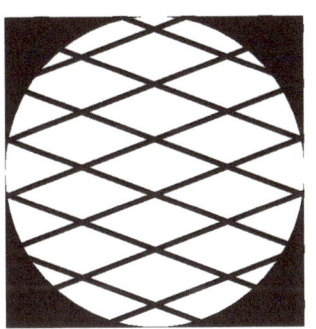

carpentered objects 9

centre–surround organisation 31–3

C.I.E. observer 19

colour matching 18–19, 42–3

colour space 42–5

complex cells 33

computational vision 12–13

cones 42–3

contrast effects 44–8

correspondence problem 26

crossed double images 26

crystal drawings 9

cubic corners 9

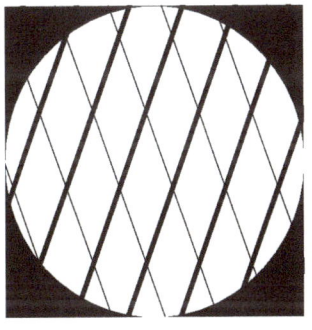

data analysis 41

dimensionality increase 48, 64

dimensionality reduction 41–2, 46

diplopia 26

disparity 26

divisive inhibition 33

dorsal pathway 14

double nail illusion 26–7

double vision 26

ecological optics 12

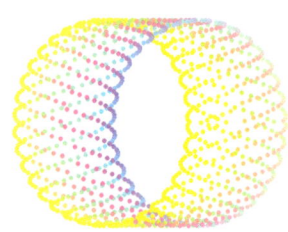

Euler angles 48–50
experience 9
eye movement space 48–51
eye movements 27–8, 36–7, 57

feedback loop 57, 59
fixation 50
fixed point 60
folded piece of card 8
forward transformation 57–9
fovea 36

geometry 17–21
Gestalt theory 11
'ghost' images 26

Hardy, Godfrey H. 15
head movements 59
Hebbian learning 42, 43
Hermann grid 31–3, 36
horopter 25
hyperplane 20

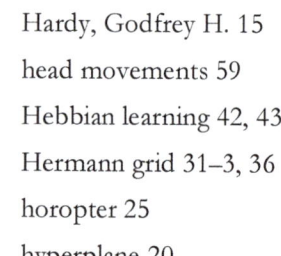

inferotemporal cortex 13
informative features 34, 37
inhibition 33
intellectual judgements 9
inverse transformation 57–9

Klein bottle 53

line of fixation 48
Listing's law 50
lms space 42
logical deduction 9

main sequence 60
manifold 41, 51, 53, 61
Marroquin pattern 35, 36
medial superior temporal cortex 14

medial temporal cortex 14
memory 37
motion parallax 21–7
motor pathway 57
motor signals 48
movement field 27
Munsell chips 41–2, 43

networks 59–60, 64
neural integrator 61
neural plasticity 43
neuron model 20–1
non-responsive region 33, 35
nystagmus 61–2

oculomotor coding 27–8
opponent colour coding 44
opponent spatial frequency 46–7
optic array 12

parallax 21–7
parietal lobe 13
perceptive field 32–3
phi movement 11
physiological optics 8–11
plasticity 43
polar diagram 23, 24
pressing on eye 8
projection 21
projective plane 53, 54

receptive field 27, 31–4
recognition 14
reflectance functions 41–2
retinal receptive fields 31

saccades 27–8, 57, 60–1
semicircular canals 59
shadows 10
silhouettes 37

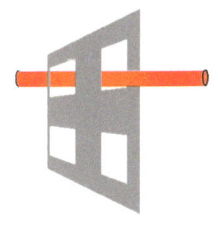

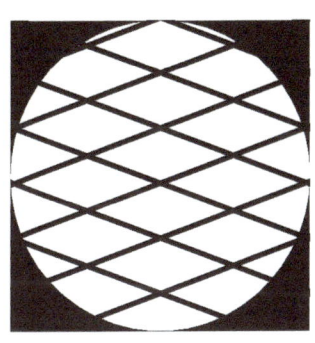

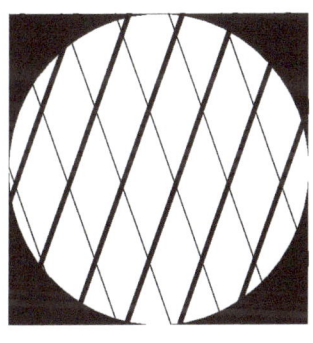

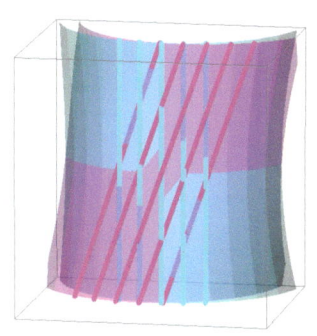

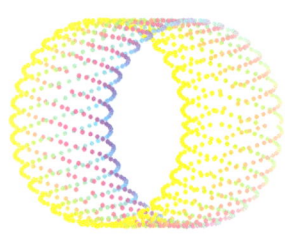

simple cells 33
simultaneous contrast effects 44–8
slow manifold 61
spatial frequency space 45–8
spatial processing 13
specific energy 11
spectral sensitivity 42–3
stereoscopic parallax 21–7
superior colliculus 27–8

Thurber, James 65
tokens 9
torus 53
trajectory 60
transformations 8, 41, 57–9
trichromatic law 18–19, 44–5

unconscious inference 9
uncrossed double images 26

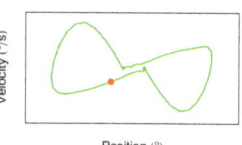

V1–4 14, 23
vectors 17–21, 23–4, 60
ventral pathway 14
visual cortex 14, 23, 33–5
visual neuroscience 14–15
visualisation 52–4

weighted inputs 20
working memory 37

NOTES

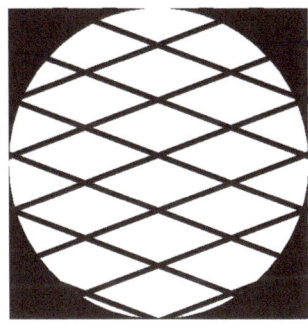

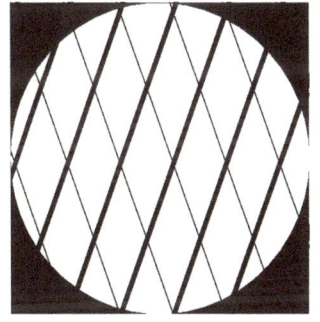

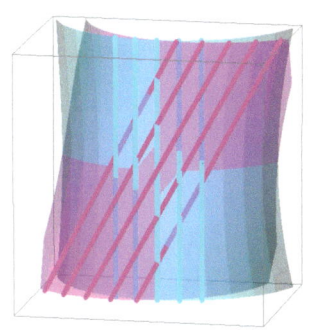

71

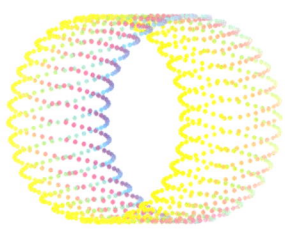

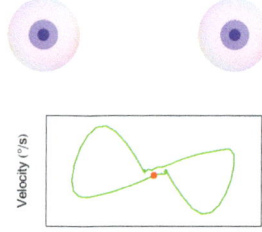

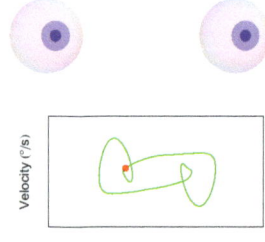

www.ingramcontent.com/pod-product-compliance
Ingram Content Group UK Ltd.
Pitfield, Milton Keynes, MK11 3LW, UK
UKHW061139180426
11947UKWH00002B/13